THE COMPLETE IDIOT'S GUIDE TO

D0792853

Urban Homesteading

by Sundari Elizabeth Kraft

HILLSBORO PUBLIC LIBRARIES
Hillsboro, OR
WITHDRAWN
Member of Washington County
COOPERATIVE LIBRARY SERVICES

A member of Penguin Group (USA) Inc.

For Brian, the best farmhand a girl could ask for.

ALPHA BOOKS

Published by the Penguin Group

Penguin Group (USA) Inc., 375 Hudson Street, New York, New York 10014, USA

Penguin Group (Canada), 90 Eglinton Avenue East, Suite 700, Toronto, Ontario M4P 2Y3, Canada (a division of Pearson Penguin Canada Inc.)

Penguin Books Ltd., 80 Strand, London WC2R 0RL, England

Penguin Ireland, 25 St. Stephen's Green, Dublin 2, Ireland (a division of Penguin Books Ltd.)

Penguin Group (Australia), 250 Camberwell Road, Camberwell, Victoria 3124, Australia (a division of Pearson Australia Group Pty. Ltd.)

Penguin Books India Pvt. Ltd., 11 Community Centre, Panchsheel Park, New Delhi—110 017, India

Penguin Group (NZ), 67 Apollo Drive, Rosedale, North Shore, Auckland 1311, New Zealand (a division of Pearson New Zealand Ltd.)

Penguin Books (South Africa) (Pty.) Ltd., 24 Sturdee Avenue, Rosebank, Johannesburg 2196, South Africa

Penguin Books Ltd., Registered Offices: 80 Strand, London WC2R 0RL, England

Copyright © 2011 by Elizabeth Kraft

The image of the drop spindles is Copyright © 2010 Pschemp, and is being reproduced with the copyright owner's permission pursuant to a GNU Free Documentation License.

All rights reserved. No part of this book shall be reproduced, stored in a retrieval system, or transmitted by any means, electronic, mechanical, photocopying, recording, or otherwise, without written permission from the publisher. No patent liability is assumed with respect to the use of the information contained herein. Although every precaution has been taken in the preparation of this book, the publisher and author assume no responsibility for errors or omissions. Neither is any liability assumed for damages resulting from the use of information contained herein. For information, address Alpha Books, 800 East 96th Street, Indianapolis, IN 46240.

THE COMPLETE IDIOT'S GUIDE TO and Design are registered trademarks of Penguin Group (USA) Inc.

International Standard Book Number: 978-1-61564-104-8
Library of Congress Catalog Card Number: 2010910073

13 12 11   8 7 6 5 4 3 2 1

Interpretation of the printing code: The rightmost number of the first series of numbers is the year of the book's printing; the rightmost number of the second series of numbers is the number of the book's printing. For example, a printing code of 11-1 shows that the first printing occurred in 2011. *5038 7553 12/12*

Printed in the United States of America

Note: This publication contains the opinions and ideas of its author. It is intended to provide helpful and informative material on the subject matter covered. It is sold with the understanding that the author and publisher are not engaged in rendering professional services in the book. If the reader requires personal assistance or advice, a competent professional should be consulted.

The author and publisher specifically disclaim any responsibility for any liability, loss, or risk, personal or otherwise, which is incurred as a consequence, directly or indirectly, of the use and application of any of the contents of this book.

Most Alpha books are available at special quantity discounts for bulk purchases for sales promotions, premiums, fund-raising, or educational use. Special books, or book excerpts, can also be created to fit specific needs.

For details, write: Special Markets, Alpha Books, 375 Hudson Street, New York, NY 10014.

Publisher: *Marie Butler-Knight*

Associate Publisher: *Mike Sanders*

Executive Managing Editor: *Billy Fields*

Senior Acquisitions Editor: *Paul Dinas*

Senior Development Editor: *Christy Wagner*

Production Editor: *Kayla Dugger*

Copy Editor: *Amy Lepore*

Cover Designer: *Rebecca Batchelor*

Book Designers: *William Thomas, Rebecca Batchelor*

Indexer: *Julie Bess*

Layout: *Ayanna Lacey*

Proofreader: *John Etchison*

Contents

Introduction

Over the last several decades, our society created a clear definition of what it meant to live in a city. As an urban resident, you could enjoy the benefits of career opportunities, cultural and social events, and endless forms of entertainment. Anything you might want could be purchased.

However, that purchasing power turned out to be a double-edged sword because buying the things that sustained you turned out to be your only option. By living in a city, you were sacrificing the ability to participate in the process of providing for yourself. Everything you used or consumed came to you from somewhere else.

Our insistence that so many things come from "somewhere else" had its consequences. The separation of the producing (out in the country) from the consuming (within the city) led to a decrease in the quality of what we eat, a propensity for waste and contamination, and a general sense of disconnection from the basic stuff of life.

What we are now discovering—or, more accurately, remembering—is that we can recapture some of what we lost and take an active role in the things that sustain us. Although we may live among skyscrapers and concrete, we don't have to be limited by them. Urban homesteaders look around and see possibilities everywhere. Where there is dirt (and even where there isn't), we can grow food. We can decide to get a chicken coop instead of a birdbath or fill a compost pile instead of a trash can. We can conserve and reuse our resources, and we can find new ways to acquire the energy and water we use to get through our day.

What it means to live in a city is changing. We still have plenty of fast cars and tall buildings, but sometimes those buildings have gardens on their roofs, and the cars have to share the road with bicycles. And some of those commuters are on their way home to milk their goats before dinner. Little by little, we are taking steps to bridge the separation between producing and consuming, and we are finding that our lives are richer because of it.

Around the world, more people now live in cities than in rural areas. We can no longer rely on those who live "out there" to bear the responsibility for producing everything for everyone, everywhere. We need to take hold of the means of production and move it toward our doorsteps. Luckily, the rewards of self-sufficient living—also known as homesteading—are more than worth the effort.

How This Book Is Organized

The elements that make up urban homesteading are notoriously diverse yet always interconnected. I've grouped the chapters of this book into five main parts:

Part 1, What It Means to Homestead in the City, introduces not only the basic concept of self-sufficient living, but also what it means to do it in an urban setting. A city provides a wealth of advantages for a homesteader, but it's not without its challenges. In these chapters, I talk about what you can do to successfully meet those challenges and why it's worth it in the long run.

Part 2, City Farming, delves into the practice that's at the core of homesteading: growing food. Urban homesteaders can get pretty creative when it comes to finding land (or concrete) to turn into a garden, and they're able to produce a surprising amount of food with just a little space. In Part 2, I talk about how to begin your plants from seed (even without a greenhouse) and protect what you grow from the weeds, pests, and city critters who might try to share in your harvest. Speaking of harvests, I also cover techniques for extending the life of your garden and getting the most productivity out of your plants.

Part 3, Raising Animals for Food, presents another aspect of food production—the contribution that can be made by small animals such as chickens and dwarf goats. These city-size livestock fit neatly into a backyard. Some others like rabbits or fish can even be kept in an apartment, while honeybees would be perfectly happy on your balcony. It's true that raising these animals in the city might require a little extra effort with the neighbors or the local zoning officials, but it is possible to integrate them into your urban homestead.

Part 4, A Homemade Life in the City, celebrates the tasks that are sometimes known as the "homesteading arts." It begins with preserving the harvest from your garden so you can enjoy your homegrown food even after the plants have died away. I also show you how you can transform the milk and meat of your homestead into other treats like cheese and stock. Expensive and chemical-laden products stand out like a sore thumb on an urban homestead, so I cover how to make homemade body products and natural solutions for cleaning your home.

Part 5, Making the Most of What You Have, acknowledges that homesteading does not always involve creating something new; sometimes it's about utilizing what's already there. In these chapters, we explore how you can meet your energy and water needs while minimizing waste and how you might transform what you discard into fertilizer for your homestead's growth. Cities are a resource-rich environment, and I talk about how you can best utilize this as you venture forth on the path of sustainable living.

Following the chapters, you'll find some helpful information in the appendixes, including a glossary and a list of resources, so you can deepen your knowledge as you continue along the homesteading path. You'll also find garden planning guides, which you can use as a reference for creating a productive garden, even in the smallest of spaces.

Extras

As you journey through this book, you'll find a few extra tidbits to help you along the way.

DEFINITION

Sometimes homesteading has a language all its own. Check these sidebars for definitions of terms and phrases you need to know to fit in with the old-timers.

SMALL STEPS

Tips for making homesteading easier, plus suggestions for starting small when approaching a new task, are highlighted in these sidebars.

ROAD BLOCK

Heed these warnings about tricky situations that could end up costing you time or money—or maybe just give you a headache!

URBAN INFO

You might find these little bits of information interesting as you venture along the urban homesteading path.

Acknowledgments

I have been blessed with a wealth of support from my local homesteading community, to whom I owe tremendous gratitude.

I would like to thank the following people for contributing their considerable wisdom to this book: Sylvia Bernstein, Ingrid Milinazzo, James Zitting, Jessica True, Tina Axelrad, Mira Gale, Cate Albert, Melanie Brooks, Katherine Cornwell, Charmaine Cheung, Amy Kalinchuk, Diane Buck, Adam Brock, Kenzie Davison, Melody Bentfield, John Carraway, and Pat Williams.

Thank you to Everett Sizemore for giving me a boost and to Shayne Madsen for lending her strength and acumen to the process.

I am grateful to Paul Dinas for guiding the creation of this book with a steady hand, to Christy Wagner for her wonderful editing, and to Katelynn Lacopo for graciously handling the peccadilloes of an inexperienced writer.

I want to thank the fantastic and inspiring urban farmers of Heirloom Gardens for keeping things going so I could devote time to this book.

And to my family and my sangha, who provide support and encouragement in more ways than I can count … thank you for everything.

Trademarks

All terms mentioned in this book that are known to be or are suspected of being trademarks or service marks have been appropriately capitalized. Alpha Books and Penguin Group (USA) Inc. cannot attest to the accuracy of this information. Use of a term in this book should not be regarded as affecting the validity of any trademark or service mark.

What It Means to Homestead in the City

Sure … it might be great to grow some of my own food. I've also heard fresh chicken eggs are delicious. I'd love to be able to can food like Grandma used to. And what about composting and using solar power instead of racking up a big electric bill? Too bad I live in the city! Oh, well … maybe someday I can learn this stuff, if I ever get out to the country.

If thoughts like these have ever crossed your mind, you've come to the right place. You *can* do all this—and more!—even if you live in the city. In fact, cities provide a lot of advantages that can actually make homesteading easier, not harder. The best part about homesteading is that you can pick and choose the elements that will fit best into your life. You don't have to go "whole hog" (if you'll pardon the expression) to benefit from urban homesteading.

What Is Urban Homesteading?

In This Chapter

- What it means to homestead in a city
- Why people choose to homestead
- Urban homesteading—an old idea that's new again
- How city living makes it easier to homestead

I went to the woods because I wished to live deliberately …

—Henry David Thoreau

And off he went, to live in solitude and tend to his bean patch. Throughout the years, many people who wished to live a life of self-sufficiency—to have a connection to the food they eat and the things that sustain them—have felt the need to follow Thoreau's lead out of the city and into less-inhabited places.

While that's all fine and good, moving to the country is not an option for everyone. What about the rest of us? What if we need to (or choose to) live in an urban area because of family, school, work, or community? Are those of us who reside in cities automatically consigned to a life of dependence on others for all of our food, energy, and water? Do we have to make a choice between a life in the city full of modern-day conveniences or the joys and struggles of self-sufficient living?

Urban homesteaders believe we don't have to choose. The distinctions between what can be done in a city and what "belongs" in the country are breaking down (and were artificial to begin with). Regardless of whether you live on an 8,000-square-foot city lot or in an apartment with only a sunny windowsill for planting space, you can live a life with all the deliberation of Thoreau.

Homesteading Defined

To get an idea of the underlying spirit of homesteading in the city, it's helpful to look back at what our homesteading ancestors were trying to do. Beginning in 1862, the U.S. government offered free land—160 acres—to anyone who could successfully live on it for five years. Determined families set out to stake their claim, far away from the support of the nearest town. In those isolated surroundings, families worked to create everything they would need to live. They grew crops, raised animals, pumped water, and handmade everything needed for the household. The family strove to live independently, to survive on what they could create for themselves.

URBAN INFO

The Homestead Act of 1862 went through a few modifications over the years, but citizens were still able to claim free chunks of federal land up until 1976 in most of the country. The exception to this is Alaska, which allowed homesteading claims until 1986.

The environment of a modern-day city is pretty much the polar opposite of what those early homesteaders faced. Instead of a life where nothing is provided for them, urban dwellers have everything they could possibly need, and then some, right under their noses. Despite this, however, some people choose instead to create things for themselves. Within a sea of fast and cheap food, people are growing vegetables in front-yard gardens and on fire escapes. Even though an endless supply of power and water can be had by flipping a switch or turning a spigot, some people are line-drying their clothes and catching rainwater in barrels.

Urban homesteading is not one specific thing; rather, it's a collection of practices that can be done while living in a city. It can include activities like growing food, canning and preserving, raising food-producing animals, making cheese and yogurt, spinning and knitting, making natural soaps and cleaning products, using solar energy, recycling household water, and composting food scraps into fertilizer. The common thread among each activity is the desire to provide for oneself, to create something rather than purchasing what is mass produced, and to treat natural resources with respect.

DEFINITION

Urban homesteading is a collection of practices, which can be done within a city, with the aim of meeting basic daily needs in a self-sufficient and sustainable way.

The list of what can be done in urban homesteading is constantly changing as modern-day pioneers find new ways to approach self-sufficient living. And as I discuss in Chapter 3, urban homesteading looks different for each person who does it.

You might have noticed in the preceding list of urban homesteading practices, as in this book as a whole, that food is a recurring theme. That's because urban homesteading is about finding self-sufficient ways to meet basic needs, and the need for food is a big one. It's also an arena where we can, with just a little bit of effort, make a real difference. By taking steps to regain some control over our personal food supply, we can live a life that's healthier and more connected to the natural world.

The Benefits of Living Sustainably

It can sometimes be difficult for friends or family to understand what's happening when someone they know gets bitten by the homesteading bug. Why go to all the trouble of digging in the dirt, cleaning chicken coops, or learning to can when anything you need is available at the corner grocery store? Once you venture into building neat little piles for your food scraps (when they can easily be thrown into the trash) or using your shower water to flush your toilet, your loved ones may think you've really gone 'round the bend.

Actually, urban homesteading is a very sensible thing to do. It provides a wealth of benefits, both for the participant and for the larger world. Everyone who homesteads is motivated by different things. One person may love making fermented vegetables not available in a store, while someone else is determined to save the planet. Regardless of the intentions behind homesteading, all the benefits still go along with it.

Quality Counts

There may be a lot of noble reasons for living a self-sufficient life, but at the end of the day, what seems to motivate homesteaders most is a desire for quality. There's just no way a tomato from the store—that was picked green, shipped hundreds or thousands of miles, and ripened using a chemical spray—could compare to the taste of an heirloom variety plucked fresh off the vine. Eggs from backyard chickens have a flavor that makes store-bought eggs seem tasteless in comparison. Creating your own soaps and lotions allows you to customize the oils and textures to exactly fit your preferences. The list could go on and on.

There's an opportunity for endless variety, too. The things available commercially are made for the masses, but all kinds of delightful possibilities exist when you have the ability to be a producer instead of merely a consumer.

Here's to Your Health

The health benefits associated with homesteading are multifaceted. First and foremost, food you raise yourself has more nutrients than its mass-produced counterpart, whether it's produce, eggs, dairy, or meat. The vitamins and minerals in produce fade the longer it sits in a box, and backyard animals eat a diverse diet that translates into a more nutrient-rich product.

Then there's the issue of everything homegrown foods *don't* have. In the case of vegetables and fruits, I'm talking about things like pesticides, chemical fertilizers, genetically modified organisms, and irradiation. Once we start looking at animal products, we're faced with bovine growth hormone, antibiotics galore, salmonella, E. coli, and all kinds of other possible contamination that can result from *factory farming*.

DEFINITION

Factory farming is a term used to describe the practice of raising large quantities of livestock in close confinement, also known as a concentrated animal feed operation (CAFO). Factory farming is the conventional, industrial model for meat, egg, and dairy production.

The health benefits of homesteading aren't limited to food. Conventional cleaning and body care products are often loaded with potentially toxic chemicals. Not only will learning how to make these items yourself reduce your exposure to harmful toxins, but it's also helpful for anyone who has a particular allergy or skin sensitivity.

And lastly, there's the benefit of the exercise you naturally get while doing the work involved in homesteading. Even though urban homesteaders very rarely get to partake in a barn raising, you can still stay fit by digging in your garden or setting up a rain barrel system.

It's About Affordability

Don't ever let anyone tell you it costs $64 to grow a tomato. There's a myth that's been built up around homesteading—that it requires a lot of costly tools and gadgets, and it's nothing but a hobby for yuppies with too much time and money on their hands. The truth is that all kinds of people (many of them with very little money) have been homesteading for generations.

It used to be simple common sense that it was cheaper to grow your own food instead of buying it at the store. Nowadays, industrial farming and agricultural subsidies artificially lower the cost of store-bought food, but it's still usually cheaper to grow your own.

The economic benefits become more pronounced when you learn ways to preserve the harvest's excesses through skills like canning and cheese-making.

URBAN INFO

Over the last decade, U.S. taxpayers have paid between $12 billion and $24 billion annually in agricultural subsidies. Almost three fourths of that money goes to the largest 10 percent of farms, and hardly any of it reaches true small family farmers. This means local, sustainably produced food (grown without the aid of subsidies) appears artificially expensive compared to what's sold in the grocery store (much of which is grown with the help of taxpayer money).

The cost savings of homesteading are even easier to see when it comes to making your own nonfood items like soaps, lotions, and cleaning supplies.

Using homesteading conservation techniques can result in a decrease in utility bills. Even options that require an initial investment pay for themselves in the long run.

Whether or not homesteading is affordable has a lot to do with your mind-set. Whenever you approach a task, ask yourself if there's a simpler or less-expensive way to get the job done—because there usually is. For example, in most cases, there's no need to build raised beds, and then buy dirt to fill them, to plant a garden. There's already dirt in the ground—try planting there!

Protecting the Environment

The litany of ways our current lifestyle is taxing the planet could fill a whole book of its own—and it has, many times. The purpose of looking at these issues is not to bog us down with guilt over how our lives are structured. But it is helpful to understand a few realities about why the way most of us live is—quite literally—unsustainable.

Consider the issue of oil, which is expensive, not just in dollars but in global politics as well. Plus, it's running out. Your use of oil is not limited to the occasions when you drive your car. In fact, that's hardly the tip of the iceberg. Oil is used in great quantities not just to grow and process your food but to transport your food to you. It's estimated that each item in an American's daily diet has traveled an average of 1,300 miles from the field to the plate. That's some serious gas.

And that's just your food. Think of all the other items we use that are shipped long distances to reach us. If we're interested in using less oil and reducing the impact of carbon emissions, then taking steps to grow our own, buy locally, reuse, and recycle becomes not just wise but imperative.

URBAN INFO

According to the Department of Agriculture, it takes 435 calories of fuel to ship a 5-calorie strawberry from California to New York.

Oil isn't the whole story, of course. Subdivisions are sprawling across formerly uninhabited regions, and there isn't enough water to support all the residents. Factory farms, which provide almost all the animal-based food in this country, are polluting the land, water, and air at an alarming rate. Toxic by-products from … well, pretty much everything that's manufactured, are contaminating the planet.

So what's a conscientious person to do? Should you throw up your hands in exasperation at the futility of it all and then go on about your business? Chuck everything and run off to live in a cave, subsisting only on what you can catch with your bare hands? There is a sensible middle ground, which can be found by incorporating urban homesteading practices into your life. Every step toward sustainability helps.

Learning Self-Sufficiency

Many people have drolly noted that humans are the only animal that cannot feed itself. We have many other skills, to be sure, but in our modern-day push for more and more convenience, we've rendered ourselves essentially helpless when it comes to the basic stuff of living. In his wonderful book *Better Off* (HarperCollins, 2005), Eric Brende shares his story of moving from a city to an Amish-like community, where he soon discovers that even his neighbor's 5-year-old son has more self-sufficiency skills than he does.

Even if our ability to live without power or grocery stores is never put to the test by a catastrophe, it's still a satisfying experience to provide for oneself and one's family. In particular, children benefit from opportunities to make a substantive contribution. Kids understand that making the bed really doesn't help anyone, but caring for the chickens that feed the family allows them to be useful in a tangible way.

SMALL STEPS

If you're looking for an easy homesteading project for kids—with fast results!—you can try sprouting. See Chapter 4 for simple sprouting instructions, and then let the kids decorate the family's salad with their homegrown greens.

Self-sufficiency—and its companion, preparedness—are often looked at as an issue of stockpiling everything one might need to survive a disaster. Urban dwellers don't have the room to store a year's worth of food and water, and doing so would require a lot of time and energy that probably can be better spent elsewhere. Preparedness doesn't have to be about what we have; it can be about what we know how to do.

Creating a Sense of Community

It might seem contradictory to say that urban homesteading allows you to become more self-sufficient while simultaneously connecting you to your community, but it's true. Self-sufficiency is not about isolation; it's about doing for oneself instead of counting on the "powers that be" to provide it for you. Sometimes the work of homesteading can be helped by lots of hands or a shared pool of knowledge, and that's where community comes in.

Community building is a natural by-product of pretty much any homesteading activity. Community gardens are an obvious example, and even personal front-yard gardens will garner the interest of neighbors. Backyard chickens and goats become an attraction for nearby families, and the animal owners gratefully turn to neighbors for help with animal care during vacations. Thrifty homesteaders can utilize time-sharing programs, which allow participants to trade their skills within a community pool. An abundance of online forums and bulletin boards allows homesteaders to tap into a larger community and learn from those with more experience.

The History of Urban Homesteading

Urban homesteading is sometimes dismissed as a fad or a trend. It's certainly true that homesteading and sustainable living practices are becoming increasingly popular, but the whole thing is hardly a new idea. In fact, it's a very old idea.

Humans have been living together in cities for centuries. And yet somehow, before the invention of highways for transporting food everywhere on refrigerated trucks, those city folk were able to eat. How do you think they managed to do it? Sure, towns and cities would have a "dry goods" store for staples like flour and coffee, but most of what people ate was grown or raised where they lived. In addition, many of the other items needed for daily life (like soaps and cloth) were either made by the families themselves or by a craftsperson in the town.

A certain level of household self-sufficiency was always the norm, but it was actively encouraged by the U.S. government during World Wars I and II. Every family was encouraged to plant a "Victory Garden" as well as raise chickens and other small livestock. Much of the food that was grown commercially needed to be shipped to the battlefields to sustain the troops, so homeowners were asked to produce as much for themselves as they could. Everyone had a garden out back, and sometimes all the available space—including front and side yards—was dug up and planted with vegetables. City dwellers raised chickens, ducks, rabbits, and goats to provide food for their families.

During World Wars I and II, the U.S. government issued posters to promote home food production.

Things started to change after the end of World War II. People began moving to the suburbs, and it became fashionable to distance oneself from the "dirty" work of growing and producing food. An expensive, impractical—but pretty!—sod lawn became a sign of affluence. Food started to become more processed, chemical-laden cleaning products promised to eradicate things like "ring around the collar," and convenience became the highest value. Life was growing more and more compartmentalized, with families residing in single-unit dwellings and living away from where they worked, went to school, and recreated. This, of course, meant they needed a car to drive pretty much everywhere.

We continued on for a while in that direction until we as a culture started to notice some problems. All that convenient, highly processed food was making us sick. So were the chemicals that seemed to be in and on everything we used. Our habit of living far away from our jobs was creating suburban sprawl with endless highway traffic jams. Our energy and water bills kept climbing, but there didn't seem to be anything to do about it.

So we started making changes, returning to a more sustainable, integrated way of living. "Mixed use" is a popular style for new developments, which allow residents to live, work, and shop within the same community. People are beginning to look around at the land that's available to them and are deciding that perhaps they'd rather grow food in their yard instead of perpetually mowing and watering the lawn. Those who live in apartments are planting vegetables in pots, on roofs, and in community gardens. People are taking steps to reuse, recycle, and make more from scratch to save money and resources.

The last half century or so was simply a pendulum swing into the realm of unsustainability, and now our society is beginning to find its way back.

URBAN INFO

The popularity of the BackYardChickens.com discussion forum is just one of the many indicators of the widespread interest in urban homesteading. The forum has more than 72,000 registered members and is growing every day.

The Advantages of Homesteading in a City

It's easy for a wannabe homesteader to harbor fantasies of living in the country, with large stretches of land for growing crops and raising animals. Many of the classic homesteading texts support that notion, with instructions for self-sufficient living that talk about space in terms of acres, not square feet. The notion of a 1-acre farm is regarded as quaint, when in fact an acre, at 43,560 square feet, is often 10 to 100 times more space than an urban homesteader has to call home.

It would be a mistake, however, to assume that more land makes for a better situation. What seems like an abundance can quickly become a burden—more to water, weed, and care for. Plus, working with just a small piece of land—whether it's a 40-square-foot community plot or a 400-square-foot backyard garden—gives you the ability to really get to know the soil and your plants and care for them in a way that larger-scale farming doesn't allow. The same concept applies to raising 6 chickens instead of 60 or making cheese from the milk of 2 goats instead of 20.

Urban homesteading also blurs the boundaries between what's literally your land and everything else that's available to you. You might do what you can to grow vegetables in pots on your balcony or raise chickens in your yard, but that's just the beginning.

Community garden (and potentially barnyard) lots, land-share garden partnerships, food-producing animal co-ops, foraging from public fruit trees, putting beehives on the roofs of local businesses ... there really is an abundance of land that can be utilized in the city—and unlike a personal homestead, you're not responsible for the mortgage!

And that's all on top of the other resources that come with city life, like salvaged building materials, endless amounts of food waste for composting, and pretty much anything gently-used-but-still-usable you can think of.

ROAD BLOCK

There's one aspect of homesteading in which rural residents do have urban folks beat, and that's the issue of zoning regulations. In Chapter 2, I talk about how you can work within—and work to change—the regulations where you live.

An additional resource urban homesteaders have in abundance is community support. This isn't to say that rural residents aren't neighborly. It's not that people in the country don't help each other out; it's just that—and this is usually the point of country life—there aren't that many people around. In cities, we live crammed next to, and frequently above and below, each other, but this can be turned into an asset. Lots of people means lots of potential hands for projects, whether it's constructing a greenhouse or building a solar water heater.

Urban farming cooperatives often have volunteers that number in the tens or hundreds because people don't have to travel far to help and are looking for an opportunity to learn. There's no way a rural farmer could reasonably expect 20 of his neighbors to show up, week in and week out, to volunteer to help him run his farm. It wouldn't happen in the country, but it does in the city.

The Least You Need to Know

- Urban homesteading is about taking steps toward self-sufficiency while living in a city.
- Whether you're motivated by a desire for good food or a wish to save the planet, you'll find what you're looking for in urban homesteading.
- Urban homesteading is not a new trend; it's a continuation of the way people lived in cities up until the last half century.
- Homesteading in a city is not a second-rate version of country living and has a number of unique advantages.

City Considerations

In This Chapter

* The rules affecting urban homesteaders
* Determining your zoning regulations
* Additional restrictions you may encounter
* Working with (or working to change) the rules

Even though homesteading has traditionally been thought of as a rural activity, you can do a tremendous amount to lead a self-sufficient life within a city, regardless of whether you live in an apartment or a house. In fact, as I discussed in Chapter 1, urban homesteaders have a better deal than their rural counterparts in several ways.

However, there's no denying that urban homesteaders often face a web of regulations unique to city life. And unfortunately, all the knowledge and desire to homestead in the world isn't going to help you if the things you want to do are prohibited where you live. In this chapter, we look at the logistical and regulatory issues urban homesteaders may face and ways you can successfully work with, or work to change, the system.

Zoning Regulations

Much of what happens structurally within cities—including private residential property—is guided by rules called *zoning regulations*. These can be tough to swallow sometimes, especially when you've plopped down a chunk of money under the impression that you own, or at least have a right to use, the space contained within your property lines. After all, why should anyone care if you keep a couple of 5-pound hens in your backyard or plant tomatoes instead of rosebushes in front of your duplex? Why is it anyone's business what you do on your own property anyway?

DEFINITION

Zoning regulations are rules created by local government regarding the location, size, structure, and function of a property (or property improvement) in a given area. The set of zoning regulations for your city is called the *zoning code*.

The unfortunate reality is that people *do* care, and zoning regulations mean they get to make it their business. Because of this, urban homesteading has a common side effect: you may learn more about your local zoning laws than you ever cared to know.

Before you contemplate civil disobedience, pause for a moment and remember that zoning regulations are meant to serve a helpful purpose. These "rules of the road" are needed to ensure everyone's space is respected whenever large numbers of people come together to share a relatively small chunk of land.

How would you feel about the "I-have-a-right-to-do-whatever-I-want-on-my-own-land" argument if your next-door neighbor was trying to raise a dozen insomniac roosters? Or if he wanted to tear down his house and build a 20-story skyscraper, dooming your garden to permanent shade? Clearly, there are times when zoning regulations are good. Of course, some rules are relevant and beneficial, and others are outdated and needlessly restrictive.

I will be addressing the zoning issues specific to different homesteading topics in the upcoming chapters. However, some basic points apply to all situations across the board.

Zoning Terms

Whether you're talking with a city employee or reading the written zoning code, you're likely to encounter a few important terms:

Accessory building A shelter for people, animals, or property that's secondary to the main building on the lot. Greenhouses and chicken coops are examples of accessory buildings.

Accessory structure Something separate from the main building on the lot. This includes all accessory buildings but can also refer to man-made structures like a fence.

Accessory use A secondary activity on your property related to the main use of the property. For example, if your property is zoned "residential," a home-based business or agricultural activities would be accessory uses.

Buffer A strip of land used to separate adjoining properties.

Grandfathering The act of allowing existing structures or practices to continue, even if they don't comply with updated zoning rules.

Green space An area covered with some kind of plant growth. This can include roofs.

Impervious surface A surface that does not permit the absorption of fluids, like rooftops or patios.

Mixed use A type of zoning or development that mixes more than one type of use on the same property or in the same building. For example, some lots are zoned for both residential and commercial uses.

ROAD BLOCK

Most cities don't allow any sort of commercial business resembling a store within a residential district. So even if you're allowed to grow food in your front yard, you would have to take it to a business district if you wanted to sell it.

Nonconforming use A use of a property that's allowed to continue, even after a new zoning ordinance prohibiting it has been established for that area. If someone was allowed to keep chickens on their lot under an older zoning code, and their right to keep chickens was grandfathered in under the new zoning code, the chickens would be a nonconforming use.

Ordinance A local law or regulation.

Primary use The main activity a lot is zoned for, like residing on a lot zoned "residential."

Right-of-way The strip of land located between the sidewalk and the street. This is also known as a treelawn, verge, parkway, or "hell strip." The right-of-way is typically owned by the city, but the resident of the property associated with it must maintain it.

Secondary use Incidental uses of a property. If a house and a garden share a lot in a residential district, the garden is considered the secondary use. See "accessory use."

Setback The minimum required distance of a structure from the property line.

Use by right A use permitted within a specified zone district that does not require special review or approval. For example, you can own a hamster as a "use by right," but you might need a permit before you can own a chicken.

Variance Permission to do something not allowed under the zoning code. For example, you would need to get a zoning variance if you wanted to build a home higher than what's allowed in your area.

Navigating Your City's Zoning Code

Zoning codes are pretty much always written by lawyers. So unless you've passed the bar exam, don't feel bad if you find the code difficult to understand. Here are a few tips you can keep in mind as you dive in and attempt to figure out what you are and are not allowed to do:

Know your designation: Start by learning what the zoning designation is for your area. Even if it's "residential," there are usually distinctions in the code between residential types (R-1, R-2, etc.). Many cities have a function somewhere on their website that allows you to input your address and receive a bunch of information on your property, including your zoning designation. Or you can call the zoning office and ask.

Search for keywords: Most cities have their zoning code online. However, sometimes it can be difficult to tell which section of the code relates to your questions. Try to find a link that opens the whole code; it will be a large document. Then use the "find" or "search" function in your web browser or PDF reader to locate all the instances of the word you're looking for, like *chicken, goat,* or *garden.*

Call and ask … and then ask again: Reading page after page of zoning language can be enough to leave you cross-eyed. Sometimes the quickest way to get answers is to call your city's zoning department and talk with someone. Be sure to make note of the name of the person who's answering your questions in case you need to reference it in the future.

But here's the catch: you should probably call back later and direct the same questions to a second person. The people who work for the zoning department certainly do their best, but zoning codes are complex and change frequently. Sometimes the first answer you get isn't the right one, so it's best to double-check.

Connect with local sustainability groups: The simplest way to understand not only what your zoning code says, but how it works in practice, is to talk with other people in your area who are urban homesteaders. They're likely to know what's in the code, what must be done to comply with it, how it's enforced, and what (if anything) is being done to change it.

SMALL STEPS

If you're not sure where to look to find local sustainability groups, try meetup. com. Whether you're interest is gardening, raising chickens, or sustainable home remodeling, you are likely to find a group of folks with similar interests in your area.

Homesteading in Apartments and Duplexes

Throughout this book I include information on the many ways you can homestead without having any land where you live, including lots of things that can be done completely within the walls of an apartment. As long as what you do doesn't create a racket (no roosters indoors, please!), you should be able to do what you'd like without bothering anyone.

You probably won't be making any major structural changes to your home like a gray water system (see Chapter 21) or a light tube (see Chapter 20) if you're renting an apartment. But if you do, it's a good idea to get written permission from the landlord first.

ROAD BLOCK

Roofs can be a great place to grow food. However, not every roof is appropriate for gardening. Be sure to read the "Creating and Caring for a Rooftop Garden" section in Chapter 4 before attempting to use your roof for homesteading.

You may live in an apartment or duplex that has some communal yard space. If you do, it's natural that you'd want to use your allotted portion for homesteading activities. For whatever project you want to tackle, even if it's something as basic as a compost pile, it's a good idea to get permission from the landlord ahead of time. It can save you headaches in the long run—especially if one of your neighbors wants to make an issue out of your activities. If you're planning a garden, you can read the "Growing in a Shared Yard" section in Chapter 5 for additional tips.

Covenant-Controlled Communities

Whereas a zoning code is a set of regulations that applies to an entire city, a covenant is an additional list of rules meant to go above and beyond what's laid out in the zoning code and that applies to a specific set of dwellings. Covenants were created with the intention of preserving property values. Most covenant-controlled communities are overseen by a homeowner's association (HOA) that enforces the rules in the covenant. Participation in the HOA is not optional; if you buy a property within the community, you are agreeing to be bound by the rules of the covenant.

SMALL STEPS

It's not uncommon for covenant rules to forbid the use of clotheslines. However, six states—Florida, Utah, Hawaii, Maine, Vermont, and Colorado—have passed some variation of a "Right to Dry" law, which stops landlords and HOAs from prohibiting clotheslines. Go to laundrylist.org/en/right-to-dry for more information.

The restrictions within covenant-controlled communities vary, but they often apply to a number of things a homesteader would be interested in. For example, there may be rules regarding landscaping, animals, accessory structures (which can include things like fences, clotheslines, or water barrels), home alterations or improvements, and the disposal of waste. It's certainly worth pausing for consideration before purchasing a property that's governed by a covenant. Urban homesteaders have enough to contend with in their zoning regulations.

But if, as happens to a lot of people, you bought your home in a covenant-controlled community before you had any interest in homesteading, and you now find yourself frustrated by the situation, there may be ways you can change things. See the "Amending Your Covenant" section later in this chapter.

Getting What You Want

So although there are numerous benefits to homesteading within a city—like an abundance of free land and resources, community support, and the manageability of small-scale projects—there are also some systemic challenges. Luckily, nothing is set in stone. Everything from a six-month lease to a zoning code can be adapted.

Plus, there's something big working in your favor that should not be underestimated. The local/healthy/organic food and sustainable-living movement is consistently gaining momentum. All across the country, individuals and communities are working to find ways to join this "green" phenomenon. Sometimes all they need is someone to point them in the right direction.

Negotiating with Landlords

A great many homesteading projects can be done without affecting your lease agreement. However, if you want to do something larger, like keeping chickens or digging up part of a yard for a garden, you have to work it out with your landlord. The best time to negotiate is often before you sign a lease. See if the landlord is amenable to the things you want to do. If he or she is not, you can look for a different place to live.

When proposing your ideas to a potential landlord, be sure to let him or her know ...

- What the project will look like.

- Where it will be located and how large it will be.

- Whether it will permanently alter the property.

- What kind of impact (noise, smell, etc.) it may have on the other tenants and your plan for dealing with problems.

- What your backup is for vacations and emergencies (if your project requires ongoing care).

- What will happen to your project when you move.

Even if you've already signed a lease, you can still ask for changes. Every situation is unique, and sometimes it will be fine to simply get verbal approval from the landlord. However, if you're considering a project that entails a significant investment of time or money on your part, it's worthwhile to get the landlord's approval in writing.

 SMALL STEPS

If you're a good tenant and are planning to renew your lease, you can use that as an opportunity to ask for changes. Responsible tenants are hard to find, and your landlord may be more amenable to your requests if it means he or she can keep you in the property.

Amending Your Covenant

Covenant rules can typically be changed through a vote of the HOA members. Sometimes a two-thirds majority is needed to make a change, and sometimes just a simple majority, such as half the members plus one, is enough.

It's unlikely you'll be able to get the change you're hoping for simply by showing up to a meeting and pleading your case. You'll need to reach out to like-minded members within your community and create a cohesive group that's supportive. The sheer momentum of it may be enough to tip the vote in your favor.

Also, HOA meetings are often poorly attended. If you can get a group of your neighbors to show up and vote, that may give you the majority you need to change the rules.

Applying for a Permit or Variance

Applying for a permit or variance can be frustrating, but it's infinitely better than living in a city that completely bans whatever it is you want to do. So while the process is bureaucratic (and sometimes expensive), at least the homesteading activity you have in mind is possible. Plus, you can always work to change your city's zoning code and remove those pesky permitting requirements (see the next section).

Part of what makes getting a permit or variance so frustrating is that the process isn't always clear. That's when reaching out to local sustainable living advocates is especially helpful because they can point you in the right direction. Keep a notebook, and document every step you take along the way of getting your permit, including the names of all the city employees you interact with. Take care when you fill out your paperwork and make it clear you're meeting the requirements they've laid out. When it comes to a permit, hanging in there through the process is most of the battle.

Requesting a variance is a slightly different story. Permits are typically applied for and granted privately, but a zoning variance usually requires public notice. This means the city gives you a sign for you to post on your home. The sign lets people know what you're applying for and tells them where they can register a comment. Generally, if too many of your neighbors—or in some cases, just one—are opposed to the variance, it won't be granted. Therefore, the process of applying for a variance needs to include a little bit of neighborhood outreach.

Your neighbors will be much more likely to support, or at least not oppose, your project if they feel they've been consulted first. Take a few minutes to touch base with your neighbors before you post your sign. It will give you a chance to give them the facts about what you're doing and address any concerns they may have. Plus, it will tip you off if one of your neighbors seems to be set against you going ahead with your project. If that person is going to register an objection against your variance, you'll want to get statements of support from your other neighbors.

Changing Your Zoning Code

If you want to take steps to make your city friendlier to sustainable living ideas like urban homesteading, that sometimes means pushing for changes in the zoning code. Some cities are operating with a code that is 50-plus years old and reflects outdated, compartmentalized thinking about the way a city should function. Part of the reason why portions of it may be so out of date is because it's not easy to change a zoning code. It's a long process that requires a lot of patience, but if you're successful, you will be creating sustainable living opportunities for your whole community.

The first step in the process is to understand who in the city government truly has the power to change the zoning code. You might be surprised to learn that it's not your mayor or the zoning department—it's your city council. They are the ones who will ultimately decide whether or not your proposed ordinance passes and becomes part of the code. And of course, they are elected officials … who generally wish to be reelected. That's why public support—and public pressure—is such a critical part of this process.

Several potential steps are involved in accomplishing a code change. Not every situation requires all these steps, but at times you'll need all the help you can get to make a change.

Make a statement: Create a clear vision of the changes you would like to see and start putting it out into the world. If you have the ability to make a simple web page, it can be a useful reference for anyone who's interested in learning more about your cause. Write a catchy e-mail that explains what you're trying to do and send it to everyone you know in your area. Ask them to help you spread the word.

Build support: You need to create a mechanism for attracting, capturing, informing, and mobilizing supporters. A Facebook page can be a great way to do this. You can also use a listserv, online message board, or e-mail mailing list.

Develop answers: Once you start talking about your ideas publicly, you'll get a sense for the kind of pushback you'll encounter. The good news is that those who oppose urban home-steading ordinances are usually not original—you're going to hear the same arguments over and over again. Put some thought into clear, memorable, and (reasonably) short answers. Then spread the answers around so any supporter who talks about the issue has a good way to respond if he or she is challenged. (If you'd like some help developing answers to common questions about urban food-producing animals, go to eatwhereUlive. com and click on "Myths Versus Facts About Food-Producing Animals.")

Find a champion: Try to identify someone on your city council who would be supportive of the changes you're seeking. Request a meeting and see if he or she would be willing to sponsor an ordinance.

Educate your community: Begin talking with community groups about your ideas and the proposed ordinance. Sustainable living organizations are great ones to focus on first because they'll be easy to talk to and will ultimately form the base of your supporters. Eventually, you want to focus your energy on your city's registered neighborhood organizations (RNOs). They have a strong position of influence in local elections, which means their opinion matters a great deal to your city council.

SMALL STEPS

All the meetings involved in a project like this can be a lot of work. Try offering training for your ordinance's supporters. Teach them how to talk about the issue before sending them to meet with their respective RNOs to advocate the cause.

Lobby your council members: Request meetings with all of your city council members to discuss your ideas for a new ordinance. If they seem to need extra convincing (usually expressed by the phrase "I need to see what my constituents think"), you can mobilize your supporters to call and send e-mails.

Create a task force: The actual writing of a proposed ordinance probably won't be done by just one person. The council member who has signed on to sponsor the ordinance can help bring together a group with the goals of researching the issue and drafting an ordinance for your city. The task force may include council members, community leaders, and people from your city's zoning department. Other agencies may also be included if it's appropriate, such as involving folks from Animal Control when writing a food-producing animals ordinance.

Gather endorsements: Once the task force has created a proposed ordinance, find community organizations willing to formally stand behind it. Groups focused on environmental issues, poverty, food justice/access, and healthy food/living are all good possibilities. Also consider asking neighborhood mothers' groups to voice their support, as well as any RNOs willing to formally stand behind the ordinance.

Continue lobbying: Throughout this process, you should be periodically mobilizing the supporters to contact their council members with their thoughts about the ordinance. They really need to know the community wants to see a change in the zoning code.

Pack the hearing room: The last step in the process is for the proposed ordinance to formally come before your city council. By the time it gets to this point, it's likely that most of the council members will have already made up their minds. However, there will still be an opportunity for community members to speak at the meeting. Do everything you can to fill that room with supporters and arrange for as many of them to speak as you can.

Clearly, it's not a simple process. Making systemic change is never easy, but it will help to improve the way people live in your city.

The Least You Need to Know

- Although urban homesteading has its advantages, some city regulations may restrict your activities.
- Talking to sustainable living advocates in your area is often the best way to learn about your city's zoning rules.
- You may have to put in the time to get a permit or variance for a homesteading activity, but it's typically worth the effort.
- If you don't like your community's rules, change them!

The Life of an Urban Homesteader

In This Chapter

- Creating your own version of homesteading
- Homesteading without disrupting the others in your home
- What you can do—some examples
- Getting a new outlook

It's a relatively simple thing to understand what urban homesteading is: gardening, raising chickens, making soaps, repurposing and recycling, …. It can be a little harder to grasp how all that might fit into your life. Everyone approaches urban homesteading with a unique equation. Your schedule, living situation, and available space all factor into what urban homesteading will look like for you.

In this chapter, we explore some of those elements, and I give you examples of what's possible when you set out to live sustainably within a city. I also spend a little time on the philosophical side of urban homesteading and share how it can impact your quality of life above and beyond the tangible benefits.

Making Homesteading Work for You

Sustainable living, especially within cities, is growing more popular every day. Sometimes it seems like you can't open a newspaper or magazine without reading about someone who's gardening, raising chickens, or using solar energy—and sometimes doing all three—within city limits.

To be sure, the increasing attention on these kinds of activities is terrific, but it can lead to a bit of a misperception as to what's required to be an urban homesteader. Would-be homesteaders might be intrigued by the idea of incorporating a few sustainable practices into their life but may feel intimidated by the prospect of upending their home à la pioneer-living-in-the-city.

Recall from Chapter 1 that I defined urban homesteading as being a "collection of practices." This is an important thing to emphasize because urban homesteading looks different for everyone. You can select the homesteading practices appropriate for your life and use them as much or as little as you want. In other words, homesteading is not an all-or-nothing proposition.

In addition, what urban homesteading means to you will continuously change. You will go through periods during which you homestead with gusto, and there will be other times when you need to order Chinese food and leave the garden untended—and that's okay.

The rhythm of urban homesteading changes naturally with the season, too. You may learn to relish the calm during winter, saving your indoor projects for those quiet months and planning for the increase in activity that accompanies the spring planting and fall harvest.

Your Schedule

There's no denying that modern life is busy. There's a lot to accomplish, a lot of places to be, but not always a lot of time to get everything done. But while there are certainly situations in which the deadlines are real, it can also be easy to get sucked into a frantic pace of our own making.

Urban homesteading provides you the valuable opportunity to step away, even briefly, from the whirlwind of city life. Spending a few minutes with your chickens as they contentedly scratch and peck has a way of washing away the stress from a busy day. Harvesting vegetables from your container garden for dinner has a grounding quality that standing in line at a grocery store just can't match.

They say you can tell what people truly value by how they spend their money. The same applies to how you spend your time. If you look at your life honestly, you can probably find time to fit in the homesteading tasks you want to be doing. Perhaps instead of spending 20 minutes each morning surfing the Internet for the latest news, you might instead spend that time listening to the news on the radio while you milk your goats.

SMALL STEPS

Don't let all this talk about schedules fool you into thinking that urban homesteading is a time-draining proposition. There are endless ways to homestead efficiently, and most projects don't take a lot of time to maintain once you get them set up.

There are times when our schedule is not under our control, like when a demanding boss or a 20-car pileup keeps us away from home longer than we meant to be. Luckily, thanks to the tricks I share throughout this book (like predator-proof chicken runs or automatic garden sprinkler timers), your homestead won't suffer in your absence.

When it comes to vacations or other extended absences, urban homesteaders often have the benefit of a readily available community to step in and help. If you're not able to find someone to volunteer for the more involved tasks like animal care, for example, you can always board your animals with a farmer in the country or pay a little money for a professional animal-sitter.

Your Companions

Your home—or homestead—may not be exclusively yours. Partners, children, and/or roommates are often part of the equation. You might be in a situation where everyone in your home is equally excited to begin homesteading, but that's often not the case. The dynamics of your home life are another element that will come into play when you start exploring urban homesteading. Fortunately, there are some things you can do to make the process smoother for everyone involved:

Start small: Change isn't easy, and a sudden or dramatic shift in your household's way of life may leave your family or roommates feeling jarred. Plus, it's often hard to tell exactly what's going to be involved in a project until you're well into it. By starting small, you'll be able to gradually incorporate homesteading practices into the fabric of your household without causing a major disruption.

Ask for their input: You may have a vision, but urban homesteading has a number of potential facets. Ask your housemates what they would like to see in terms of sustainable-living practices in your home. You may discover that although your partner isn't really interested in gardening, he or she thinks gray water systems are the neatest thing since sliced bread.

 URBAN INFO

Sharon Astyk wrote a wonderful essay about the common phenomenon in which one person becomes enraptured with homesteading while their partner is left wondering what on earth has happened. You can find a link to her essay, "So You (Don't Particularly) Want to Be a Farmer," in Appendix B.

Don't require involvement: This is a bit of a judgment call. The argument certainly can be made for using urban homesteading as a way to teach kids responsibility. It's also reasonable to ask your partner to give you some support as a basic requirement of … you know … partnership. However, it's not really fair to drag your loved ones (or roommates) into the work of homesteading if it's your passion and not theirs.

Plus, it might backfire. Your housemates aren't likely to see the charm of owning chickens if they're regularly cleaning the coop against their will. If you know you're the only one in your household who is interested in urban homesteading, choose projects you can do largely on your own.

Share the benefits: Despite all the noble intentions for urban homesteading, the primary motivator is often simple enjoyment. The products of homesteading—like fresh organic vegetables and eggs, handmade soaps, etc.—are typically of a significantly higher quality than what you've been purchasing at the store. Be generous with your family or room-mates, and they will quickly warm to the idea of urban homesteading. Once your partner gets hooked on goat milk yogurt or fresh honeycomb, helping out won't seem like such a chore!

Your Money

Frugality and economic sustainability are core principles of urban homesteading. This way of living is grounded in the notion of providing for oneself, so buying lots of "stuff" in order to homestead would be contrary to its basic purpose. You shouldn't be spending a ton of money to be an urban homesteader. In fact, you will almost certainly save money in the long run.

Many projects do require a small initial investment, like purchasing a couple of tools for your garden or building a chicken coop (although you can minimize the cost of a coop by building it with repurposed materials). Once you get your homesteading projects up and running, you have the potential to save money on food, cleaning supplies, body care products, energy, and water. Plus, if you engage in bartering or time-banking programs, you can receive a wide variety of goods and services without spending any money at all.

URBAN INFO

It can sometimes be difficult to convince people they'll save money on food by growing it themselves when a hamburger can be purchased for 99 cents. The truth, of course, is that there are layers of hidden costs that come along with that hamburger. Industrial, mass-produced food creates significant societal costs related to health care and environmental cleanup and repairs. You don't pay for these things when you're ordering at the food counter, but we all pay for them in the long run.

A Look at What's Possible

"There's no use trying," Alice said. "One can't believe impossible things."

"I daresay you haven't had much practice," said the Queen. "Why, sometimes I've believed as many as six impossible things before breakfast."

—from *Alice in Wonderland*, by Lewis Carroll

There are endless potential versions of an urban homestead, but we're going to look at a couple as examples of what can be achieved while living in a city. These hypothetical "days in the life" might contain more than you're interested in doing, but they do represent a sliver of what's possible. You'll find more information on everything described here (plus lots more) later in this book.

With a Backyard

If you're a homesteader with a small backyard, your day might begin with a visit to the chicken coop. You gather a few eggs from the nesting boxes and pause in the small garden planted next to the chicken coop to pick a tomato. You're eating fresh tomatoes every day because they're ripening so quickly right now, and you'll probably can a batch of spaghetti sauce over the weekend. As you look at your tall tomato plant, you can remember the day you started it indoors from a seed, as well as the day you transferred that small seedling out into the garden.

While you cook a quick breakfast of scrambled eggs with diced tomato, you sweeten your tea with honey from your neighbor's bees. He keeps a couple of beehives but no chickens, so you periodically trade him some of your extra eggs for honey. He loves the eggs so much he gives you a little beeswax, too, which you use to make a simple three-ingredient lotion.

When you get home from work, you spend 10 minutes pulling weeds out of the garden, which you give to the chickens, who happily gobble them up. The chickens eat not only the weeds but pretty much all your food scraps. Once a week or so you rake the straw and chicken manure out of their run and add it to your compost pile. The weather has been dry lately, so you open the spigot on the barrel you use to collect rainwater during storms, which now flows into your garden's drip hose.

SMALL STEPS

Fresh chicken manure is too strong for plants, but if you have some that's been aged for a while, you can add it directly to the garden (and skip the composting step).

You head inside and open your utility bill, which shows, yet again, that you don't owe anything for electricity. The solar panel on your house generates enough energy during the day to run your meter backward, so your energy use in the evenings is already paid for.

Speaking of things that use energy, your computer has been acting strangely lately. You don't need to worry about shelling out money to get it repaired, though. You're going to spend an hour this weekend tutoring someone in your time-banking program and cash in the credit you earn to get computer repair from another member of the program. Thank goodness for bartering!

In an Apartment

What if you're an urban homesteader without any land? Your day might begin with a smoothie made from homemade goat milk yogurt and foraged peaches. You participate in an urban goat co-op, so twice a week, you ride your bike a few minutes across town to take your turn milking a couple of backyard goats. In exchange for your participation, you get to take home jars of fresh milk. Some of the milk stays in liquid form (to be poured on cereal and in coffee), and some of it becomes yogurt and soft goat cheese.

SMALL STEPS

Although yogurt- and cheese-making might sound complicated, the process is pretty simple. It typically involves adding some kind of culture to the milk and then setting it aside for a few hours while the bacteria in the culture do most of the work.

You gathered the peaches this weekend during an afternoon foraging trip with a local sustainability group, picking from local trees whose owners have offered them up to be harvested.

During your morning shower, you use a bar of homemade soap, which also contains your goat milk. You collect some of the water from your morning shower in a bucket, which you then use to water your houseplants and windowsill garden.

On your way home from work, you stop by your community garden plot to harvest kale, carrots, and lettuce. You're planning on making a soup for dinner, but there are so many carrots, you gather a few extra. Before leaving, you wave to the other gardeners, who you'll be seeing this weekend during the community garden's group work day.

When you arrive home, you grab a few mushrooms from the box where they're growing in your closet and snip some fresh herbs from the pots in your windowsill. The last time you roasted a chicken, you made stock from the leftovers and froze it in ice cube trays. Now you can use a few of the frozen stock cubes to add flavor to your soup, along with a handful of dried tomatoes from last year's garden. While your soup is simmering on the stove, you make a salad with lettuce from the garden, sprouts growing in the jars on your counter, and a vinaigrette made with your windowsill herbs.

The extra carrots from the garden go into your crock of lacto-fermented vegetables. You gather the trimmings from your dinner preparation and add them to your worm bin so the worms can compost the scraps into the fertilizer that feeds your garden plot.

Finally, your roommate is able to clean the kitchen after dinner without aggravating her chemical sensitivities because she's using homemade cleaning products made with natural ingredients like vinegar and baking soda.

A New Sense of Connection

> *We discovered that [the chickens] loved to eat the weeds I pulled, and the grasshoppers I caught red-handed eating my peppers. ... Growing hens alongside my vegetables, and hornworms and pigweeds as part of the plan, has drawn me more deeply into the organic cycle of my gardening that is its own fascinating reward.*

—Barbara Kingsolver, "Lily's Chickens," *Small Wonder*

There's no denying that urban homesteading yields tangible, measurable benefits. You can do it simply for the good food or the financial savings, and that's fine. However, many who set out to live sustainably are also looking for something bigger—for a way of connecting to the natural world, the greater community, and the things that sustain them.

Our modern society is built largely around principles of comfort and instant gratification. Those may sound like beneficial things (after all, who doesn't want to be comfortable?), but they come at a price. We exist in a cocoon of climate-controlled homes, cars, and offices. We're insulated from shifts in the weather, and we don't tend to notice precipitation, or lack thereof, unless it affects our commute. Modern-day grocery stores show no evidence of the seasons—everything from tomatoes to winter squash is available every day of the year. A dazzling array of prepackaged, precooked food is always accessible, ready to be hastily consumed as we dash from one obligation to the next.

This is not to say that comfort, and even instant gratification, doesn't have its place. Even the most committed urban homesteader wants to stay warm during the winter and send out for pizza from time to time. However, there's real value in reintegrating oneself with the rhythms of the natural world. Gardeners develop a deep connection with the seasons—the rush of growth in spring, the gaudy extravagance of summer, the slowing down of autumn, and the stillness of winter. As the year draws to a close, this year's plants become next year's compost, and the cycle begins again.

We begin to see our food not as a commodity but as a process. We are able to recognize the time and natural resources that went into producing what's on our plate, and we cherish our meals more because of it. We value every drop of rain we collect for our garden and each ray of sun we harness for energy. We do what we can to create an integrated system in which nothing goes to waste and where each part supports the growth of the whole. We begin to expand this sense of connection beyond the boundaries of our own property, and we work to contribute to our larger community and protect our global environment.

The Least You Need to Know

- You can homestead a little or a lot—it doesn't have to be all or nothing in order to be valuable.
- Regardless of your schedule, family situation, or finances, you can find ways to fit homesteading into your life.
- It's best not to force the other people in your household to join you in homesteading, although once they experience the benefits, they may decide it's worthwhile.
- Don't underestimate what you can accomplish as an urban homesteader.

City Farming

Gardening is usually the gateway activity to urban homesteading. At the heart of homesteading is self-sufficiency, and tending to a patch of vegetables is a great way to feel the pride that comes from providing for yourself and your family. Cities offer lots of ways to grow food, whether you own land or not.

Urban farmers may be dealing with space in terms of square feet instead of acres, but there are ways to make even the tiniest garden surprisingly productive. You don't need to limit yourself to just vegetables as a city farmer, either—you can plant fruit trees, grapevines, and a berry patch, too!

In this part, I cover the many different ways to grow food in the city, plus I give you the tools you need to tackle potential garden problems and have a successful harvest!

Growing Without a Yard

In This Chapter

- Scouting out good places for growing food
- Growing vegetables and herbs in containers
- Gardening on rooftops
- Raising healthy sprouts
- Simple mushroom gardening

The idea of what a garden can look like has changed tremendously, thanks to the growth of urban homesteading. For those who don't have a yard, or don't choose to grow in it, gardens are sprouting up in all sorts of places—on patios, on roofs, on countertops, and in closets. There are many advantages to growing food in this way. Plants raised outside of the ground generally have fewer problems with weeds, pests, and diseases. They're also easier to tend to, making them ideal for anyone who may have trouble crouching or bending.

The ability to move your garden can be a great benefit, too. Many in-ground gardeners would love the option of moving plants out of the path of oncoming hail. In addition, growing a garden that's not in the ground gives you the capability to take everything with you if you move out of your home in the middle of the summer.

In this chapter, we look at the many places you have available, both inside and outside, to grow vegetables. I describe the steps for successfully gardening in containers and look at ways to expand your garden to your rooftop. We also explore additional ways to produce food in small spaces indoors, through sprouting and mushroom growing.

Finding Good Growing Spaces

Wherever there is sunlight, a plant can grow. If you keep this simple rule in mind, you're likely to find lots of potential growing spaces in your home. Windowsills, kitchen counters, or floor spaces near windows are all good possibilities. You'll find that south-facing windows provide the most light. East- and west-facing windows can also work well, but north-facing windows may not give you the sunshine you're looking for. It varies somewhat from plant to plant, as shown in the following table, but in general, plants need around six hours of sunlight daily to grow well.

Sunshine Needs	Vegetables and Herbs
Less sunshine	Lettuce, spinach, arugula, chives, cilantro, parsley, *perennial* herbs (oregano, rosemary, tarragon, sage, thyme)
Moderate sunshine	Carrots, beets, turnips, broccoli, kale, chard, peas, radishes
More sunshine	Tomatoes, peppers, eggplant, summer and winter squash, melons, cucumbers, basil

DEFINITION

A **perennial** plant is one that lives for more than two years. The plant may appear to die or stop growing in the winter, but it will return the following year.

Once you venture outside the walls of your apartment or house, you may find lots of additional space for growing. A balcony, porch, stoop, or patio can all be wonderful spaces for a container garden. (I talk about rooftop growing later in this chapter.)

Certain plants, like cucumbers and small winter squash, can be trellised to grow vertically, so you can look at walls and railings as potential growing spaces. Pots can also be hung from hooks, so your ceiling or patio overhang can be used as well. Floors, walls, ceilings—pretty much any surface you see can be used for your container garden!

Keep in mind the amount of sun a space receives and know that more is not always better. A certain amount of sun is necessary for growth, but unremitting sunshine for hours and hours a day is likely to cook even the heartiest plant.

Container garden vegetables and herbs don't have quite the same buffers against the elements (sun, wind, rain) in-ground plants have, so in that regard, they are more delicate. The nice thing about a container garden is that you have complete control over where the pots are located, so you can find spots with the right balance of sunshine and shade.

SMALL STEPS

If it's your first time setting up a container garden, or your first time in a new location, you can begin by putting small pots of vegetables in a wide variety of places inside and outside your home. This way, you can discover which places get the best sunlight and then focus your subsequent plantings in those areas.

Small Container Gardening

When you think of a container garden, you may remember your grandmother keeping a few herbs in a windowsill planter, or maybe you envision growing a small tomato plant in a pot on the back patio. The truth is, the sky's the limit when it comes to what can be grown in a container garden. Some plants do require bigger pots than others, and some vegetables may be better suited to container growing, but ultimately, you should feel free to plant whatever you want.

If it's your first time trying container gardening, herbs are a wonderful way to begin. They're small, easy to care for, and fun to harvest and use for cooking. Some people recommend focusing container gardening efforts on the items that are the most expensive to purchase—like tomatoes, gourmet salad greens, and fresh herbs. However, don't worry too much about what you "should" plant in your containers. Ultimately, if you plant what you like to eat, you'll enjoy your garden.

Choosing the Right Containers

Almost any container can be used for growing vegetables as long as it provides good drainage. It should have holes in the bottom (or along the side near the bottom) that are ½ inch in diameter. Containers are often made of plastic, terracotta, glazed ceramic, or wood:

Unique containers can be used for any kind of vegetable, but potato growing seems to bring out the especially creative side of gardeners. It's not uncommon to find people growing potatoes in plastic tubs, wooden barrels, or even old trash cans!

Hanging baskets can be a lovely way to grow vegetables and herbs, and they help maximize growing space by moving a portion of your garden up into the air.

Growbags are another way to garden without planting in the ground. Growbags are pretty much what they sound like—bags you fill with soil and then plant with vegetables. They're often made from peat or polypropylene, materials that "breathe" and allow drainage for the plants. Growbags have the advantage of being lightweight and easy to store in the winter.

Although you can grow pretty much anything you'd like in a container, certain varieties of each vegetable fare better in pots. Often these are dwarf or compact varieties, or they may be types that are sturdier and need less support.

Knowing the minimum container size required for different crops can come in handy when working with limited space. The following table lists container sizes for popular crop varieties.

Recommended Container Sizes and Crop Varieties

Crop	Minimum Container Size	Suggested Varieties
Beans	2 gallon	Bush varieties like Blue Lake, Contender, Dragon Tongue*
Beets	2 gallon	Early Wonder, Chioggia, Golden
Broccoli	2 gallon	Waltham 29, Bonanza
Carrots	2 gallon	Little Finger, Danver's Half Long, Thumbelina
Cucumbers	1 gallon	Lemon, Spacemaster, Salad Bush
Eggplant	2 gallon	Listada de Gandia, Bambino, Rosa Bianca
Herbs	½ gallon	Any
Kale	2 gallon	Blue Curled Scotch, Russian Red
Lettuce	1 gallon	Little Gem, Salad Bowl, Lollo Rossa, Tom Thumb
Onion	2 gallon	Scallions, Tokyo Long White, Crimson Forest
Peas	2 gallon	Sugar Snap, Oregon Sugar Pod
Peppers	2 gallon	Red Mini Bell Pepper, Goat Horn Hot, Jalapeño
Potatoes	5 gallon	Red Thumb Fingerling, Yukon Gold
Radishes	1 gallon	Perfecto, Early Scarlet Globe
Spinach	1 gallon	Bloomsdale Long Standing, New Zealand
Squash, summer	3 gallon	White Scallop, Eight Ball
Squash, winter	3 gallon	Buttercup, Acorn
Swiss chard	1 gallon	Any
Tomatoes	3 gallon	Sweet 100, Yellow Pear, Early Girl, Cherokee Purple
Turnips	2 gallon	Purple Top White Globe, Golden Globe

Pole bean varieties work well if you'd like to trellis the plants.

Remember that the more space a plant's roots have to grow, the healthier (and larger) the plant will be. So if your space and budget allows it, opt for bigger containers when possible.

Before filling your clean containers with potting soil, line the base of each container with newspaper to prevent soil loss. Then, fill them within an inch of the top with damp potting soil. You can follow the planting instructions outlined in Chapter 5 regarding sowing seeds and transplanting seedlings. Be sure to set your containers on bricks or blocks to allow proper drainage.

The Dirt on Potting Soil

You have many different possibilities to consider when deciding what you will use to fill your pots. One option that's *not* recommended is filling your containers with soil straight from the ground. It's typically too dense and compact for container gardening, and it could contain weed seeds and plant disease organisms.

You can purchase commercial potting soil from your local garden center. It's available in soil-based and soil-less varieties. It's good to experiment a little to find the growing medium that works best for you, depending on the climate where you live. Soil mixtures tend to hold water better, and soil-less mixtures drain faster.

Purchasing potting mixture in the volumes required for container gardening can get expensive, so some people choose to create their own. There are probably as many potting soil recipes as there are gardeners. If you'd like to mix your own potting soil, you can try the following recipe. All of the ingredients should be available from a local garden supply center. If you have trouble finding coconut coir at a store near you, you can find it online.

DIY Potting Soil

$\frac{1}{3}$ sifted compost

$\frac{1}{3}$ coconut coir

$\frac{1}{3}$ vermiculite, perlite, or builder's sand

Blend ingredients in a bucket until well mixed.

URBAN INFO

Peat moss used to be a common ingredient in homemade potting soil mixes. It has fallen out of favor in recent years because the process used to harvest it is considered environmentally destructive. Coconut coir is a renewable resource, made from fibers found between the inner and outer shells of a coconut, that is now often used in place of peat moss.

Caring for Your Containers

Container growing has many advantages. However, plants grown in pots are more vulnerable to extremes in weather compared to plants grown in the ground. They're also more at risk of drying out quickly because they have less soil from which to draw water. Exposure to wind or long periods of direct sun will speed up the drying out of the container soil and increase the need for watering. During hot periods, plants need to be watered daily (and sometimes twice a day).

You can test to see if the container needs watering by putting your finger in the soil. If the soil sticks to your finger, the container probably doesn't need to be watered. Be careful not to let the soil dry out completely between waterings because that can damage the plant's roots. More is not always better, though—overwatering can deprive the roots of oxygen and wash soil nutrients out the bottom of the container.

When watering, add water until the potting soil is thoroughly damp. Try not to get the leaves of the plant wet when watering because wet leaves can lead to plant diseases. If you'd like some help regulating the soil moisture, you can add an inch or so of mulch on the top of the potting soil. Straw, grass clippings, pine needles, and old leaves are all good mulch materials.

ROAD BLOCK

If you are keeping your containers on a balcony, be sure the water doesn't run out of your pots onto your neighbors below. Some containers come equipped with a basin to catch water, or you can set your containers on a tray filled with small rocks.

If your soil regularly dries out, you can try grouping your containers together so the foliage creates a canopy to shade the soil. If your schedule allows it, you can also take advantage of one of the great benefits of container gardening—portability! Position your containers in the sun for the recommended time and then move them into the shade to give them some respite from the heat. Planting in light-colored containers can also help lessen heat absorption.

Container plants need more regular fertilizing than their in-ground counterparts. Purchasing potting soil that contains organic fertilizer, or including organic fertilizer or compost in your homemade potting soil, is a good place to start. A great way to ensure consistent nutrients for your container garden is to use a liquid fertilizer once a week when watering. Try not to get too enthusiastic with the fertilizer, though. In this instance, as with watering and sunlight, more is not better. Container plants have less soil to buffer them from extremes, so adding too much fertilizer can damage the plants or create a nutrient imbalance that results in poor vegetable formation.

Once your plants begin growing, it's time to start thinking about trellising. Tomatoes (particularly *indeterminate* varieties), peas, and pole beans all need some type of support. Peppers and eggplants don't always need support, although sometimes it's beneficial for plants with large fruits. Cucumbers and small winter squash don't have to be trellised, but you can do so if you'd like to save space. You can use cages, stakes, or trellises for your container plants.

DEFINITION

Determinate tomato varieties ripen all of their crop at once and are typically bushy and compact plants. **Indeterminate** tomatoes continue to set and ripen fruit throughout the course of the season until the plant is killed by frost. Indeterminate plants are larger and grow in vines, which require more support.

Assisting your garden in growing vertically not only helps make the most of your space and supports healthy plant growth, but it can be beautiful as well. Pea vines can be easily trained to climb a trellis and form a beautiful green wall or even to grow across an arbor to create a canopy.

Getting the Most Out of Your Containers

It's possible to maximize your available container space by planting multiple vegetables in a single pot. You can use the spacing recommendations in Appendix C to decide if what you're envisioning for your container will work. For example, a round pot with an 18-inch diameter could comfortably hold 4 pepper plants.

It's also possible to plant different varieties of vegetables in a single container. Salad greens are tasty and easy to grow, and you can easily create a "salad pot" by planting spinach, arugula, lettuce, and endive seeds in the same container! The companion planting recommendations (see Appendix C) can be helpful in determining which plants grow well together.

It's also good to consider the root depth of the various vegetables. A deeply rooted plant like a tomato in the center of the pot, surrounded by shallowly rooted lettuce plants, is a good combination.

URBAN INFO

We think of tomatoes, peppers, and eggplants as part of our vegetable garden, so why are the products of these plants called fruits? Technically, each of these (along with squashes, cucumbers, and even beans) is a fruit. I'll skip past the technical botanical jargon and give you an easy way to understand it: if the part of the plant you eat contains seeds, it's a fruit.

It's possible to have two or three successive plantings in the same pot in a season, especially if you're growing vegetables that mature quickly. You'll want to freshen up the potting soil between plantings. Empty the pot and mix in fresh potting soil—about one part new soil per three parts old soil.

While your pot is empty, scrub it with a little liquid dish soap and water. However, if the previous planting failed because of pests or disease, you'll want to be more vigilant. In that case, discard all the old potting soil and clean the pot with a bleach solution (about 1 tablespoon bleach per cup of water).

Rooftop Gardening

Urban gardeners who wish to grow beyond their balcony or patio have the option of expanding their garden onto their rooftops. There's a developing movement of planting greenery on roofs in cities because of the multitude of benefits it provides. Extra gardening space for urban dwellers is just the beginning.

Rooftop gardens deflect heat and keep buildings cooler, saving energy that would otherwise be spent on air conditioning. The energy savings equates to less air pollution, and the plants growing on rooftops help filter the air. Rooftop greenery absorbs rainfall and reduces urban runoff that would otherwise collect toxins and empty into sewers.

Creating and Caring for a Rooftop Garden

Rooftop gardens can take many forms. The simplest option is to create a container garden with lightweight, moveable pots. You can also construct raised garden beds that sit on the surface of the roof, but the beds must be carefully constructed to protect the roof from water. This involves a multilayered system that directs excess water to the roof's existing drainage system, and utilizing the assistance of a professional rooftop garden installer is recommended.

Before beginning a rooftop garden, it's important to have a licensed professional (a structural engineer or an architect) determine how much weight your roof can support, which will tell you what kind of garden you can build. The perimeter of the roof is generally the strongest part, so keep that in mind if you'll be using containers for your garden. You'll also need to assess the condition of the roof, including whether it's safe to be on and accessible for garden maintenance. Unless you want to spend a lot of energy carrying water up to the garden daily, you'll need to have access to a spigot or a water collection system up there, too.

URBAN INFO

Bay Localize (baylocalize.org) did an assessment of what would be possible if all the rooftops in Oakland, California, were converted to gardens. It estimated that these gardens could grow 124 metric tons of leafy green and deep yellow vegetables each year, supplying the recommended intake for 8,500 residents.

The same conditions that exist for patio container gardens apply to rooftop gardens, only more so. Sun exposure can be unforgiving on rooftops, and wind speed doubles for every 10 stories of building height. You'll likely want to build your garden in a spot that's protected from the wind or construct some sort of windbreak that will also provide partial shade.

It's especially important with a rooftop garden to be sure your plant containers are not leaking water, which could pool on the roof and increase the weight load on the building. Speaking of weight, it's best to utilize lighter materials when choosing or constructing containers for a rooftop garden.

Zoning and Permitting Issues

If you're renting your house or apartment, you need to consult with your landlord before proceeding with plans to start a garden on the roof. If the rooftop is a shared area, it's also wise to talk to the other tenants in the building. You may find that your neighbors are excited about the project and want to assist in the garden's creation.

The next step is to check your city's zoning code to be sure there are no prohibitions against rooftop gardens. If your garden involves construction, you'll need to go to your city's building department and follow the steps to receive a building permit.

Micro-Gardening with Sprouts

Sprouting can turn your kitchen counter into one of your most prolific gardening spaces. Sprouting is the act of soaking and rinsing seeds, beans, nuts, or grains until they germinate. The resulting sprout is delicious and nutritious. Sprouts can be used in salads, in stir-fries, and on sandwiches. Sprouted beans can be puréed to make a spread, and sprouted grains can be ground and used for bread.

Unless you live in a warm climate, traditional gardening can be difficult during the winter months. Sadly, this is also the time when the quality of vegetables and fruits available in stores decreases (while the price increases). Sprouting can allow you to enjoy fresh, affordable, and homegrown vegetables daily, even during the coldest part of winter.

Any number of seeds, beans, nuts, and grains can be used for sprouting. Here are some popular choices:

- *Seeds:* alfalfa, clover, radish, broccoli, mustard, sunflower
- *Beans:* mung, lentil, garbanzo, pea (Kidney bean sprouts are toxic. Do not use them to sprout.)
- *Nuts:* almonds, filberts (a.k.a. hazelnuts)
- *Grains:* wheat berries, rye, quinoa

The technique for sprouting is straightforward and requires a minimum of equipment. A quart canning jar, cheesecloth, rubber band, and a small tray or dish—plus your seeds and some water—are all you need. It's important that all of your equipment is clean, and your sprouting jar should be sterilized. Sterilize the jar by running it through the dishwasher, immersing it in very hot water, or soaking it in a bleach solution.

Here are the basic steps for sprouting:

1. Measure your seeds, between 2 tablespoons (for small seeds) and 1 cup (for large beans and grains). Inspect the seeds, and remove any that are withered or split.

2. Rinse the seeds thoroughly, and place them in a sterilized jar. Add water to the jar until the water level reaches around 2 inches above the seeds. Let seeds soak for around 8 hours (up to 12 hours for large beans or nuts).

3. Secure cheesecloth to the mouth of the jar with a rubber band to make a filter, and drain the water. Rinse the seeds by filling the jar with water again and pouring it out. Shake the jar to remove as much water as possible. Place the jar upside down in a tray or dish, tilted at a slight angle to allow any remaining water to drain out.

4. Leave the jar on a counter for 3 to 5 days. Almost every kind of sprout prefers to be grown in indirect sunlight. The exception is mung beans, which need to grow in the dark so they don't become bitter.

5. Repeat the rinsing twice a day (in the morning and evening). If you live in a warm climate, you may want to rinse the seeds three or four times a day to keep them from souring.

6. Taste the sprouts as they grow and use them whenever you like the flavor. Most seeds' sprouts (like alfalfa and radish) are best when they reach 1 to 1½ inches, lentils usually only grow to ½ to 1 inch, and mung sprouts are 2 or 3 inches when they're done.

7. Store finished sprouts in the refrigerator.

Seeds grown in a clean jar and thoroughly rinsed and drained should sprout without any problems. If the seeds are not drained properly, they may rot and should be discarded. Sometimes sprouts develop mold; these should also be thrown away.

On the other hand, don't allow the seeds to dry out, or the sprouts will die. Sprouts left to grow longer will develop leaves and become baby greens. Allowing sunflower seeds to sprout for 7 to 10 days is a popular way to grow baby greens.

If you'd like to grow multiple batches of sprouts in one container, consider purchasing a multitiered sprouter.

Growing Mushrooms

If your patio has been converted to a container garden, you have herbs growing on your windowsill, and sprouts are germinating on the kitchen counter, you may think you have nowhere left to grow food. Think again. And think mushrooms.

The beauty of growing mushrooms is they prefer to be in the dark, so your closet may be the perfect place to raise a crop of shiitakes. Growing mushrooms can be an art form, and many experts are dedicated to propagating exotic mushroom strains and creating custom compost blends in which to grow them. For the beginner, though, kits are readily available that make growing your own mushrooms fast and easy.

You'll find a lot of variability among the mushroom kits you see online. Searching for "mushroom farm in a box" will lead you to the most user-friendly kits, which are available in white button, portabella, and shiitake varieties. If you're willing to venture past premade kits, you can also find the spawn (started mushroom growth, similar to purchasing a vegetable seedling instead of seeds) for many additional types of mushrooms.

Be aware that different mushrooms prefer different growing mediums, so if you're buying the spawn alone, you need to search out the appropriate medium to go with them.

 ROAD BLOCK

If your mushroom kit already contains soil, it will be presterilized for you. However, if you're purchasing spores and providing your own growing medium, it's important that it be sterilized first. Bake the soil in a 200°F oven for 1 hour. This prevents the introduction of strange spores that could grow into mushrooms that are not safe for eating.

Basic mushroom-growing kits require you to keep the soil damp and store the mushroom box in a cool, dark place. That's it. The mushrooms will be ready to harvest in about a month and will continue to grow for three to five weeks.

If you've ordered spawn and are putting together your own growing box, you'll place the sterilized growing medium in a non-pressure-treated wooden box, sprinkle in the mushroom spawn, and cover the spawn with an additional 2 inches of sterilized medium. Water the box, cover with a damp cloth, and continue to moisten the soil and cloth as needed.

After three or four weeks, you'll see that the spawn have rooted and sent little white threads across the growing medium. Remove the damp cloth and cover the growing spawn with an inch of potting soil (no need to sterilize the potting soil). Lightly water the soil and the cloth, cover, and continue to moisten as needed. Within about 10 days, the mushrooms will begin to form, and 10 days after that, they should be ready for harvesting!

The Least You Need to Know

- All kinds of vegetables and herbs successfully grow in containers.
- Windowsills, patios, balconies, and rooftops are all potential sites for a container garden.
- Growing sprouts is an easy way to produce your own vegetables all year, using very little space.
- Even the dark places in your home, like closets, can be used for gardening if you grow mushrooms.

Growing in Your Yard

In This Chapter

- Transforming your yard into a garden
- Building good soil for healthy plants
- Planting techniques for high productivity
- Addressing zoning, legal, and neighborhood concerns

Some city dwellers have at least a little open land where they live. Whether it's a small square of dirt by the front stoop, a strip of land along the path to the parking area, or a 400-square-foot front yard, it can be used to grow a garden. Even if the available area is tiny, there are ways to produce lots of food by using small-space planting techniques. A wide range of vegetables, herbs, and fruit can be grown in the city, right in your back—or front—yard.

In this chapter, I go over the basics of converting all or a part of your yard into a garden. Good soil is a big part of a good garden, so I talk about creating healthy soil, too. I give you some options for a garden layout, along with ways to plant so you can grow lots of vegetables and fruit in a small space. I close the chapter with a discussion of the zoning and neighborhood challenges that sometimes come up when gardening in the city.

Yard-to-Garden Conversion

After a winter spent reading gardening books and ordering seeds, dreaming of the time it'll be warm enough to plant again, it can be tempting to go a little overboard when planting your garden in the spring. However, it's important that you create a garden that's manageable for you. If gardening begins to feel like work, there's a good chance garden maintenance will fall by the wayside before autumn.

Just remember that you can always expand the garden next year and again the year after that. The "right" garden size is different for each person or family. Some gardeners are served well by a 40-square-foot space, and other people begin with 200 square feet. But there's one bit of advice that's fairly reliable: if it's your first time gardening, this is likely not the year to tear up your entire yard. Start small, discover what works for you, and go from there!

Choosing a Location for Your Garden

It may be that you have just one small patch of land available to grow in. If so, your decision about where to plant a garden becomes easy! As long as the space gets at least six hours of sun each day and has access to water, you can begin planting. However, if you have multiple potential garden spots, you'll need to choose the best place (or places). Remember that gardens don't have to be located in backyards. The soil beside or in front of your home is just as garden-able!

ROAD BLOCK

It's very important to locate the utility lines before you begin digging in your yard—for your safety and for the integrity of the utilities. Dial 811 at least 48 hours before you begin digging to receive free utility locating services. It's required by law in most cities.

Here are a few factors you'll want to keep in mind when choosing the location for your garden:

Sun exposure: Most plants need around six hours of sunlight to thrive. Some like more, and some can get by with a little less. (See the table in the "Finding Good Growing Spaces" section of Chapter 4.) Generally speaking, it's not a good idea to plant a garden on the north side of a building or tall fence. Be aware of trees that are in the vicinity of your garden. They may not look like much in the middle of winter, but when their branches are full of leaves in the summer, they could end up shading your garden.

Water access: Unless you live in an area with lots of reliable rainfall, you need to have a way to water your garden. Even if you believe Mother Nature will provide all the moisture you need, it's not a bad idea to have a backup plan. There are many ways to water a garden, including in-ground sprinkler systems, drip lines, oscillating sprinklers, and hand watering. Proper plant spacing (which I discuss later in this chapter) helps limit evaporation and conserve water.

Good drainage: There's an old saying that plants don't like having wet feet. This just means that plants with roots in consistently water-logged soil won't grow well and are more prone to disease. Constructing raised beds (see the "Growing in Beds" section later in this chapter) can help with soil drainage, but it's a good idea to plan your garden in a well-drained area if possible.

Soil quality and impediments: The soil in your garden supports the plants and provides them with nutrients. Since the health of your soil is so important, I discuss tips for improving your soil later in the chapter. Soil depth and impediments are also worthy of consideration. If possible, try digging a little in your potential garden spot to be sure you don't hit rock or concrete when you're half a shovel down. Some urban sites have a layer of soil that's so compacted, it might as well be concrete. Nearby trees may give you additional challenges if their roots reach into your garden space.

Protection from pets: Your dog may be your best friend, but your garden is one place where she shouldn't have free rein. After you've spent time lovingly planting your garden, the last thing you want is to worry about your dog stepping on plants and digging things up. Plus, the feces from dogs (as well as cats) is toxic to humans, and it's not safe to have pets going to the bathroom in your garden. This is especially important if anyone who is pregnant, elderly, or immunocompromised will be eating the vegetables.

Security: Putting a garden in your front or side yard has many advantages, but it's not without its risks. Most of the people who pass by your garden will admire it from afar, but it is possible that someone may steal vegetables or walk over and damage your plants. You're the best judge as to whether a front-yard garden space would be respected in your neighborhood. If not, you may decide the garden would do better in your backyard.

Visibility: Your best bet for staying caught up with your gardening tasks is to plant your garden in a place that's easily—and frequently—visible to you. You'll be more likely to notice when weeds or pests are getting out of control, and you may even find yourself taking an extra minute to tend to the garden whenever you pass by.

Growing in a Shared Yard

If your potential growing space is in a yard you share with others, you'll want to take a few extra steps to ensure the growing season goes smoothly. It's a good idea to talk with the other residents of your building about your plans. Even if they don't technically have veto power over the garden, they'll appreciate knowing what's going on before you start digging.

Plus, this will give you a chance to be sure nothing else has been planned for the space you're interested in. Few things are more discouraging than doing all the work to prepare a garden space and then being told you can't plant there because one of your neighbors has already claimed the spot for his birdbath.

If the water for your building is a shared expense, you'll need to work that out with the other residents (or the landlord) before you begin your garden.

ROAD BLOCK

If you are renting your home, be sure to check with your landlord before transforming the yard in any way. Odds are your landlord will be supportive of your garden plans, but if not, you could be stuck with the expense of relandscaping your garden area when it comes time for you to move.

Creating a Garden Space

There are two main ways to convert a yard into a useable garden space, whether it consists of sod, bare dirt, or lots of weeds. One way is to remove the existing landscaping and dig *soil amendments* into the garden. The other option is to pile soil amendments on the garden space, burying the existing landscape and building new garden soil. People do, of course, sometimes take a little from each approach and mix and match to create a system that works for them. You may run into gardeners who are fervent advocates of one method or another, but each has its advantages and drawbacks.

The first method of creating a garden is fairly straightforward. Begin by digging up and removing any weeds and undesirable plants growing in your garden space. Remove sod with a shovel or rent a sod-cutting machine from a local hardware store. Next, spread soil amendments over the surface of the garden area. You will need to mix the soil amendments into the dirt, either by turning the ground over with a shovel or by using a tiller.

A variation on this method is a technique known as double-digging. It's best done after you've already marked your garden beds because it's best to double-dig only the areas you plan on planting. To double-dig a garden bed, spread a 1-inch layer of compost over the surface of the bed. Dig a trench 1 foot wide and 1 foot deep across the width of the bed. As you're digging, place the soil you remove into a wheelbarrow or bucket. After you've dug the 1-foot trench, use a garden fork to loosen the soil in the excavated area.

Repeat this procedure by digging a second trench next to the first—only, instead of putting the removed soil into the wheelbarrow, use it to backfill the first trench. Continue digging in this way until you reach the end of the bed, where you'll need to use the dirt from the wheelbarrow to fill the last trench. Double-digging is labor intensive, so it's a good idea to start small. Try double-digging one bed the first year, and you can do subsequent beds in following seasons.

If you'd prefer not to dig, you can create your garden by putting a lot of soil amendments on top of your future garden space. This technique is sometimes called *lasagna gardening*.

It gets its name from the layers you build. Begin with a layer of cardboard or several sheets of newspaper and wet this material to help keep it in place. Next, alternate layers of *browns* (like leaves, shredded newspaper, straw, or coconut coir) and *greens* (like aged manure, fruit and vegetable scraps, coffee grounds, or grass clippings). You want your brown layers to be approximately twice as deep as your green layers.

DEFINITION

A **soil amendment** is any material added to the soil to improve its physical properties, resulting in better plant growth and health. Soil amendments can improve the soil's structure and provide additional nutrients for the plants. Commonly used soil amendments include compost, aged manure, grass clippings, and leaf mold.

Browns is a term for composting materials that are high in carbon. **Greens** refers to materials high in nitrogen.

Continue layering until you have 18 to 24 inches of soil amendments on top of your garden space and moisten everything periodically with water from the garden hose. Over time, the amendments will compost and shrink down, and eventually, you can plant your garden right in the layers you've created.

If you choose to use the first method and dig soil amendments into your garden space, you can create a garden in a short period of time with a relatively small amount of amendments. Sometimes it's not possible to plan your garden months in advance, and if you use this technique, you can plant just a little while after you start digging.

However, digging—particularly if it involves sod removal—can be a lot of work. Also, some people believe that digging (and especially tilling) causes damage to the soil structure and increases erosion. This problem is more severe in situations where farmers are tilling their fields multiple times a year as opposed to a home gardener who is likely only digging or tilling up the garden once a year. Nevertheless, many gardeners choose not to dig or till.

The lasagna gardening method has the advantage of bypassing the work of digging the garden. Whatever is in the ground now—whether it's weeds or grass—will be mulched and, over time, become part of the garden soil. Also, by layering browns and greens, you can create wonderfully fertile ground in which to plant.

However, lasagna gardening requires a significant amount of outside inputs. It can be challenging to locate and transport home enough material to create an 18-inch covering of soil amendments. If you plan on using your household's wastes, it's difficult to find enough room to store everything before building the garden. It takes time for the layers of soil amendments to break down, so you would need to start this project in the fall if you hope to plant in the spring. Be aware that no matter how carefully you construct your layers, your neighbors may consider the old leaves and food scraps in your yard unsightly. Lastly, it is possible for your soil amendments to blow away in a strong wind, especially before they've had a chance to start composting.

Creating Healthy Soil

The soil is the heart of your garden. It is the single most important factor in growing strong and healthy plants. In fact, some gardeners spend more time working to "grow soil" than anything else. It is a process that takes time, but there are some things you can do right away to make improvements.

Everyone who begins a garden starts with a patch of dirt that has its own issues. Your soil may have too much sand, too much clay, or too many rocks. It may be compacted after decades of being walked on, or it may be depleted of nutrients because of whatever was previously growing there. Luckily, with a little effort—and a healthy dose of compost!—you can begin to correct these problems.

ROAD BLOCK

Beware of overfertilizing. Although it's tempting to think that if some is good, more is better, that's not always true. For example, tomato plants that have too much nitrogen will grow to be huge and lush, but they won't produce very many tomatoes.

Soil generally needs to be improved in two ways—structure and nutrients. Soil structure is the way soil granules bind together. Gardeners generally have either soil that's sandy and doesn't hold water well or soil that's clay and difficult to dig. Ground that's become too compact is a problem for plants, whose roots need to be able to easily penetrate the soil.

The ideal soil structure is light and crumbly with good moisture retention. To improve a garden's soil structure—whatever the problem—add organic materials. Compost is the undeniable queen of soil amendments, but there are other options as well. Anything I mentioned earlier as a possible ingredient for lasagna gardening (like leaves or aged manure) can be added to improve soil structure. Some combination of browns and greens is the best way to go.

The second aspect of soil health is nutrients. Veteran gardeners can diagnose soil deficiencies by watching how the plants grow, but for everyone else, it can be difficult to know which nutrients are missing. The best way to check your soil's health is to get a simple soil test. Your state agriculture extension office can mail you a kit, and it's relatively inexpensive. Testing also tells you the pH (degree of acidity or alkalinity) of your garden. The ideal soil pH varies with each vegetable, but most plants do well with a pH between 6.0 and 7.0.

If Your Soil Has …	Add …
Low pH	Lime or limestone
High pH	Sulfur
Low nitrogen	Aged manure, soybean meal, cottonseed meal
Low phosphorus	Rock phosphate, bone meal
Low potassium	Greensand, wood ash, rock potash, seaweed meal
Low calcium	Gypsum, eggshells

ROAD BLOCK

Aged manure is like gold to a gardener, but it can also be hazardous if the "aged" part is missing from the equation. Manure you purchase from a store will already be aged, and if you buy from a local farmer, ask for the older stuff. To be safe, only use manure that's at least 6 months old.

Getting a soil test for a new garden space isn't always mandatory. If you'd prefer, you can plant your garden, add some general soil amendments, and it's likely that everything will grow well. Soil testing is simply a way to gather more information. However, if you have reason to think the soil you'll be growing in could be contaminated, you should contact your agriculture extension office and follow its recommendations for soil testing. See Chapter 6's "Determining If the Land Is Safe" section for additional information.

Designing a Garden in the City

Gone are the days when a vegetable garden always meant tidy little rows of plants separated by large areas of bare dirt. Of course, you can still plant that way if you wish, but today's urban gardeners are utilizing a wealth of other options. Whether you're most interested in productivity, beauty, resource conservation, or ease of care—or maybe all of these!—you can design a garden to fit your needs.

One way to look at an urban vegetable garden, especially if it's planted in a front yard, is to think of it as "edible landscaping." You landscape as you wish with shrubs and flowers, so why not throw some vegetables into the mix? You can fully landscape your yard with edible plants or just add a few vegetables here and there to augment what you already have growing.

URBAN INFO

For centuries, Europeans planted tomatoes as ornamentals in their gardens, but they never ate the fruits. Tomatoes were thought to be poisonous.

Growing in Beds

At the most basic level, there are two ways to organize the vegetables in your garden. The first would be to plant in single rows, with walking paths between each row of vegetables. The other way is to create beds (or large patches) of vegetables evenly spaced throughout the bed, with no space within the vegetables for walking. It still needs to be possible to navigate through the garden, of course.

Bed gardens consist of wider beds of approximately 3 or 4 feet, bordered by narrower paths of approximately 1½ to 2 feet. It's important to be able to reach halfway into the garden beds from each side so you never have to step into the beds to tend to the plants.

There are a multitude of reasons why a gardener—especially when growing in a small space—would choose to design a garden with beds instead of rows. Beds are a much more efficient use of space because the ratio of planted dirt to bare dirt is significantly higher. A garden designed with beds is up to four times more productive than the same area planted in rows.

Bed gardens also have fewer problems with weeds and soil erosion, and they require less water. Bare dirt, which makes up a majority of the garden if it is planted in rows, attracts weeds, whereas the gridlike plant spacing in a bed serves as a living mulch to help keep out weeds. The plants in a garden bed use their leaves as a canopy to shade the soil and prevent evaporation, lessening the need to water. Soil erosion can more easily occur in the multiple walking paths of a row garden; the higher plant density in a bed garden helps prevent erosion.

A final benefit is improved soil structure. A commonly observed rule of bed gardening is that, once the bed is formed, it's never walked on. This allows the soil to stay light and aerated, resulting in healthier plants.

A common way to design a bed garden is by building raised beds. Raised beds should be 3 or 4 feet wide and can be whatever length fits the garden. The recommended height for raised beds is 6 to 12 inches, but if you have difficulty bending down, you can raise the beds to at least 22 inches tall.

Raised beds can be constructed out of cedar, cement blocks, bricks, untreated wood, rocks, and even straw bales. (Avoid using pressure-treated wood or tires because these can release toxic chemicals into the soil.) Each potential raised bed material varies in terms of its costs relative to its durability. The best bet for low-cost, durable building materials is to locate used concrete blocks or bricks. You may be able to get these at a discount from a supplier if you request damaged ones, or you can scout nearby construction sites for old concrete.

SMALL STEPS

Freecycle (freecycle.org) and Craigslist (craigslist.org) are both wonderful resources for locating materials to build raised beds. Someone in your area might be doing some remodeling and trying to get rid of exactly what you're looking for!

It's also possible to create a raised bed without using any materials at all. Using a shovel, mound your garden soil and fashion a bed with the width, length, and height you're looking for. This will naturally erode over time, so it will require more maintenance than a bed built with materials.

Raised bed gardening has become incredibly popular, and it offers a number of advantages. The soil in raised beds heats up faster than soil in the ground, and it also drains better. The soil in raised beds is often not taken from the ground but added in, so the soil medium can be composed of ideal materials and will start out free of weeds and diseases. The soil won't suffer from compaction because the beds are never walked on. Raised beds offer a way for people who have difficulty bending, such as the disabled or elderly, to enjoy gardening. And if your soil is contaminated, raised beds may be a way for you to grow food safely.

Raised beds are not without their problems, though. They can be very expensive, especially when constructed out of a naturally decay-resistant wood like cedar. Purchasing compost and soil to fill the beds is also costly. Building raised beds is time-consuming and labor-intensive, at least to gardeners who aren't handy with tools. If you're gardening in a dry area, the improved drainage of a raised bed isn't a plus. In fact, Native Americans in the Southwest used to plant in "bent beds" that sloped *into* the ground to retain moisture. Raised beds may require more soil amendments over time because you're growing in the same soil year after year. You also lose the flexibility to easily redesign your garden and try new layouts.

A way to sidestep these challenges is to design your garden with in-ground beds. A basic layout might consist of 4-foot-wide beds alternating with 2-foot-wide paths; simply mark the corners of the beds with stakes, garden flags, or small rocks. Once the beds have been marked and dug, don't walk on them.

An additional bonus of in-ground beds is the ability to *cover crop* your paths and shift the garden beds each year. Cover crops are plants that improve the soil where they grow, mulch out weeds, and help the soil retain moisture. Farmers use a number of cover crops (see "Preventing Weeds" in Chapter 8), but Dutch white clover is a good choice for planting garden paths. When the garden layout rotates, the cover cropped paths will help keep the soil from wearing out from season to season.

DEFINITION

A **cover crop** is a plant grown primarily to add nutrients to the soil. Cover crops are typically tilled or dug into the ground after a specific period so they can be incorporated into the soil.

Permaculture

Permaculture is a contraction of the words *permanent* and *agriculture*, and it refers to the creation of a system that imitates the relationships found in nature. The goal is to design a space—in this case, a garden—where plants are able to flourish with a minimum of effort from the gardener.

The first step in permaculture is to observe what's naturally occurring in your future garden space. You may see sunny spots, shady spots, places where the ground dips (and collects water), or places that are higher up (and get less water). If your garden is bordered by a wall that gets a lot of sun, you have an especially hot spot. You can utilize these naturally occurring sections (sunny/shady, wet/dry) to decide which vegetables to plant where based on what they need to thrive.

Plants can also be partnered in mutually beneficial ways. For example, beans like to climb, and lettuce does well in partial shade, so you could create a trellising structure for beans in front of the lettuce bed.

Permaculture is not a set of specific techniques, but a way of working as effectively as possible with the naturally occurring system.

Getting the Most Out of Your Garden

Small space gardens can be tremendously productive. In fact, square foot to square foot, small gardens have a level of productivity a farmer could only dream of! By using the techniques I discuss in this section, it's possible to grow all the following plants in a 256-square-foot garden. (See the garden planning guides in Appendix C for an example.)

8 tomato plants	38 perennial herb plants
25 pepper plants	263 arugula
12 eggplant plants	526 spinach
10 summer squash plants	172 beets
6 winter squash plants	86 turnips
172 pea plants	150 carrots
108 kale	416 scallions
208 lettuce	383 radish sprouts

Of course, these numbers assume every single thing that's planted will germinate and grow, which every gardener knows isn't likely. However, even if three fourths of what was planted in the garden survived, it would be quite a harvest from a small space!

Getting this kind of mega-harvest out of a mini-space is wonderful, but it doesn't have to happen the first time you plant a garden. Trying to follow *all* the guidelines of companion and succession planting can be a little overwhelming to a new gardener. It can make creating a garden map feel like trying to juggle too many balls at once! It's okay to keep things simple when you're starting. Focus on plant spacing in your first year, and incorporate aspects of companion and succession planting in subsequent years.

Proper Plant Spacing

Plant spacing is a bit of a balancing act. Plants that are too far apart waste precious garden space. Plants that are too close together inhibit each other's growth and won't be as productive. The ideal plant spacing gives each plant's root system and leaves adequate room to grow—just that much and no more. For many plants, the ideal spacing is one in which the leaves of neighboring plants barely touch when the plants reach maturity. The leaves form a canopy over the soil, reducing available space for weeds and helping keep the soil moist. (See Appendix C for plant spacing recommendations.)

Planting vegetables at designated intervals is a relatively simple task when transplanting seedlings like tomatoes or eggplant into the garden. It becomes more difficult to achieve proper spacing when planting seeds. Larger seeds, such as beans, peas, chard, and beets, can be spaced by using a planting frame. Smaller seeds like carrots, lettuce, spinach, and radishes are best scattered over the soil. Spread the seeds with the ideal plant spacing in mind; later on you can "thin" the seedlings (pull some of them out) to create the proper spacing.

SMALL STEPS

It's easy to build your own planting frame. Start with chicken wire or plastic fencing material with 1- or 2-inch holes. Using four pieces of wood (furring strips work well), create a wooden rectangle as wide as your garden bed and 1 or 2 feet long. Cut the fencing material in the shape of the wooden rectangle and attach it to the frame with a staple gun or double-point staple nails. Place the frame down on your garden soil and use the spaces as guides when plant-ing. For example, if you have 2-inch fencing material and you're planting peas in a 4-inch spacing, put a pea into every other space in the planting frame.

Companion Planting

The plants in your garden affect each other. When placed nearby, some plants can assist each other to grow well by depositing needed nutrients in the soil and repelling insects. For example, onions or garlic can be an effective pest-deterrent when interplanted among other vegetables. Conversely, other plants may have the opposite effect.

Good plant companions also come into play when deciding which vegetable follows what in succession planting (see the next section). For example, planning a first planting of peas (which leave nitrogen in the soil) followed by tomatoes (which use nitrogen to grow) is beneficial. (See Appendix C for companion planting suggestions.)

Gardeners don't always know why companion planting works, but it's been successfully practiced for centuries. Native Americans developed a planting group known as the three sisters—corn, beans, and squash. The corn provides a structure for the beans to climb, eliminating the need for trellising. The beans add nitrogen to the soil, which boosts the growth of the corn and squash. While corn and beans grow up, squash grows out, shad-ing the soil and acting as a living mulch to block out weeds.

While companion planting is still more art than science, it is helpful to know each plant's "loves" and "hates" when planning your garden.

Succession Planting

Productive gardens are not planted all at once. In fact, some sections of the garden can be planted two, three, or even four times during a growing season. To accomplish this without wearing out the soil, it's important to keep companion planting guidelines in mind. Also, try not to plant vegetables from the same family in succession. For example, if you begin the season by planting broccoli in one of your beds, it's best not to follow with arugula after the broccoli is harvested since they're both in the Cole family.

Other factors important in succession planting include how early the vegetable can be planted and how long it typically takes to produce food. Some vegetables can go outside as soon as the soil has thawed enough to be "worked" (which means you can stick a shovel in it) in the spring. Other plants are more sensitive and can't be put into the ground until after the *last frost date*.

DEFINITION

The **last frost date** is the approximate date for the last frost of spring. The first frost date is the approximate date for the first frost of the fall. Many plants, primarily the nightshade and squash families, can only survive outside between the last and first frost dates. To find out your area's frost dates, check with your agriculture extension office.

Seed packets generally list the "days to maturity," which is supposed to be how long it takes the vegetable to be ready for harvest. The truth is that a day in April is not the same as a day in July, nor are the days in Maine and Arizona comparable. In other words, days to maturity can vary considerably.

It's more helpful, for the purposes of succession planting, to think of vegetables in terms of being short-, medium-, or long-maturing types. These categories just refer to the general amount of time it takes for the plants to mature.

Depending on the length of your growing season, you may be able to plant three (or even four) rounds of short-maturing vegetables into one bed. Gardeners can sometimes fit two rounds of medium-maturing vegetables into a bed during a season. A long-maturing vegetable may rein over one bed for the entire summer, but it might be possible to sneak in one harvest of a short-maturing vegetable beforehand. For example, tomatoes take a long time to mature, but you could seed a quick lettuce crop in the bed during the spring before the tomatoes are planted.

See Appendix C for the recommended planting dates and time to maturity for different vegetables. If you'd like to learn how to extend your harvest past the first frost, see Chapter 9.

SMALL STEPS

One way to extend a garden's harvest is to find tricks for planting cold-sensitive crops a little earlier. A Wall O'Water is a ring of plastic that can be filled with water and placed around individual plants, protecting them from freezing. They're available at garden centers and can be ordered online. Some gardens make their own version by encircling the plant with water-filled 2-liter plastic soda bottles.

The Fruits of Your Labor

Urban gardens need not consist only of vegetables and herbs. You can add berries, grapes, and even apples or peaches to your harvests. One of the best things about fruit plants and trees is that they're perennials, so they don't need to be replanted each year. In fact, your patch of strawberries or raspberries will actually expand on its own, creating new plants every year. It can be a relief, after a spring spent putting in a vegetable garden, to have some plants around that reliably produce food without much work from you.

If you'd like to plant a fruit tree in your city yard, select a dwarf variety. These trees only need an 8-foot-diameter space in which to grow, and their smaller size makes tree care and harvesting easier to accomplish. The exact size of each type of dwarf tree varies, but in general, they're only half as tall (around 12 feet) as a standard fruit tree. But even though the trees themselves are smaller, they still produce regular-size fruit. Apple, cherry, peach, plum, pear, nectarine, apricot, and citrus trees are all available in dwarf sizes. Most dwarf trees begin bearing fruit within three to five years.

The best way to choose a specific variety is to visit a local nursery and ask which trees do well in your climate. If you're only planting one tree, look for a *multigraft* type that will *pollinate* itself. If you'd like a smaller fruit tree that's even easier to manage, look for "mini" dwarf trees—you can grow them in pots!

DEFINITION

A **multigraft** tree has three or four compatible varieties grafted (attached) to a single tree, so the tree will self-pollinate. **Pollination** is the transferring of pollen from one flower to another so the plant will reproduce (which, in this case, means it will bear fruit).

If a fruit tree doesn't fit into your garden design, or if you'd like to start harvesting fruit sooner rather than later, you have a number of options:

Strawberries: These are easily the most beloved fruit of gardeners. Strawberries are available in June-bearing (which produces one crop in late spring or early summer), everbearing (produces two crops during the season), and day-neutral (produces fruit continuously throughout the growing season) varieties. Strawberry plants send out runners, which root in the soil to create new plants. Homegrown strawberries do tend to be quite a bit smaller than the ones available in stores, but the flavor is much better.

Raspberries and blackberries: These plants are called brambles, and they produce somewhat prickly canes on which the berries grow. Some varieties don't produce fruit until the second year, but everbearing (raspberries) and primocane (blackberries) varieties will produce the first year. Raspberries and blackberries spread so well some consider them invasive. They'll take over a garden space if they're not kept in check!

Blueberries: These little fruits grow on shrubs that are so attractive they're sometimes grown just as ornamentals. Blueberry shrubs produce small white flowers in the spring, lovely greenery in the summer, and leaves that turn red in the fall. Blueberries are available in lowbush (smaller fruits that are often called "wild" blueberries, sometimes grown as groundcover) and highbush (larger fruits, upright shrubs) varieties. Blueberries need acidic soil (pH of 4.5 to 5.2) to grow. If the soil in your area is not acidic enough, it is possible to grow blueberries in pots by using a growing medium of 40 percent coconut coir, 40 percent peat, and 20 percent perlite.

Grapes: Grapes need to be trellised, and it often works well to grow them along a fence. You can choose to grow "table grapes" which are suitable for eating, or "wine grapes," which are good for making wine. Grapes can be raised in most environments, but it's best to visit your local garden center to learn which specific varieties grow best in your area. It typically takes around three years for a new grapevine to produce grapes.

Zoning, Legal, and Neighbor Issues

Some cities have restrictions on what can be grown in front yards. Even if you call your garden "edible landscaping," the city may not be fond of front-yard tomatoes. Sometimes the limitations on gardens take the form of height restrictions for front-yard vegetation. Other cities have a ban on what they call "row crops," which is not a clearly defined term, so you may need to ask for clarification.

If you live in a covenant-controlled community, it's likely you'll need to get permission before planting anything edible in your front yard. Also, the right-of-way may be subject to zoning regulations (and HOA restrictions) of its own. The right-of-way is technically owned by the city, even though it's the property owner's responsibility to maintain it. Because the city owns the space, they have the right to dig it up if they need to access something (like a water line). So if you plant in the right-of-way, understand that there's a chance the vegetables could end up with their roots in the air!

Gardening in general is getting more and more popular, and front-yard gardens are becoming commonplace. Nonetheless, your neighbors might not be too happy about a vegetable garden taking root next to their Kentucky bluegrass lawn. If you take good care of the space, and maybe share a little of the harvest with your neighbors, it's likely they'll come around in time.

Most front-yard gardeners find that neighborhood residents become very interested in the garden and take any opportunity they can to talk with the gardener about the vegetables' progress. In fact, planting a front-yard garden is a tried-and-true way to get to know your neighbors better.

The Least You Need to Know

- When selecting a location for your garden, look for good sun exposure.
- Planting vegetables closely together helps reduce weeds and water evaporation and allows you to grow more food in a smaller space.
- Fruit trees can be a part of your urban garden, but berries will produce fruit for you more quickly.
- Check your local zoning ordinances (and your neighborhood covenants, if applicable) before planting in your front yard or right-of-way.

Growing on Someone Else's Land

In This Chapter

- Sources of free (or cheap) land in the city
- Determining if land will make a good garden
- Creating successful land-sharing partnerships
- Great aboveground gardens

Land space is at a premium in many cities. Apartment dwellers enjoy many benefits of their choice of home, but large yards are not one of them. Some people live in complexes with shared yard space that can't be gardened, and others reside in covenant-controlled communities that dictate the landscaping down to each individual blade of grass. What's an aspiring urban farmer to do?

Luckily, cities contain all kinds of available land, if you know where to look for it. In this chapter, I discuss various low-cost ways to create a garden near where you live. I talk about what you should be aware of when looking for a new garden space, and go over the zoning issues involved. Plus, we explore some of the techniques for creating gardens on concrete, so you'll have even more options when looking for a place to grow food.

Community Gardens

Community gardens are a popular way to grow food for those who don't have a garden at their home. Created and maintained by the people who live in the neighborhood, community gardens are increasingly becoming a part of the urban landscape. In fact, according to the American Community Gardening Association (ACGA), around 18,000 community gardens are growing throughout the United States and Canada. With manageable plot sizes and the benefits of community support, city gardeners can experience all the joys—and challenges!—of growing food for themselves.

DEFINITION

A **community garden** is a large piece of land (often a whole city lot) that's divided into small garden plots.

Community gardens play a multifaceted role in their neighborhoods. Some examples of community garden core values are to …

- Provide access to healthy food.

- Promote healthy lifestyles.

- Donate a portion of the harvest.

- Teach sustainable practices, like composting and water conservation.

- Serve as a source of pride for the gardeners and the neighborhood.

Participating in a Community Garden

If this sounds like something you'd be interested in, the first step is to locate a community garden in your area. Searching online for "*(your city)* community garden" or "*(your city)* urban garden" is probably the easiest way to start. You can also use the community garden finder on the ACGA website (communitygarden.org). Many gardens have waiting lists, so it's important to sign up for a garden plot as early as possible. Sometimes this means planning ahead and signing up in the fall!

Community garden agreements vary from place to place, so be sure to ask for all the details before you commit to a plot. Plots can range considerably in size—anywhere from 16 to 150 square feet, depending on the garden. Although that may not sound like a lot of space, you'd be surprised at how much food you can produce! (See "Getting the Most Out of Your Garden" in Chapter 5 for information on how to plant small spaces for maximum productivity.)

If you want more space, it's possible your community garden may allow you to rent two plots, or you could join forces with the gardener of a neighboring plot. You can coordinate your plantings, share the harvest, and give your vegetables a little more room to spread out.

ROAD BLOCK

Unfortunately, vegetable theft can occur in community gardens. Some gardeners are able to shrug off missing vegetables as part of the urban gardening experience, but it can be upsetting. If keeping all your vegetables in place is important to you, look for a garden with a good fence.

Gardeners are asked to pay a small fee for the year, typically in the range of $30 to $50. Most gardens offer assistance for participants who are unable to pay. This fee goes toward covering the cost of the water. Some gardens have in-ground sprinklers or drip systems, but often gardeners need to hand water. The garden organization is responsible for maintaining the spigots, fences, and lights. Each gardener needs to bring their own seeds, organic fertilizer, and gardening tools.

If you're not sure you'll have the time (or know-how) to manage a garden plot alone, share it with a friend! Having someone to weed with makes the time go faster, and it's nice to have a backup when life gets busy and you're not able to get to the garden when you'd like to.

Many community gardens have group work days when gardeners can come together to tend to the paths, perimeter, and other shared areas of the gardens. This is an opportunity to experience the "community" aspect of community gardens, when gardening newbies get a chance to pull weeds next to seasoned growers. It's a good idea to talk to other local gardeners when you can because you may learn valuable tips you won't find in gardening books. The soil, rainfall, weeds, diseases, and pests you'll encounter are all specific to where you live. In community gardens, the advice is often as abundant as the vegetables!

Starting a Community Garden

Gardens, scholars say, are the first sign of commitment to a community. When people plant corn they are saying, let's stay here. And by their connection to the land, they are connected to one another.

—Ann Raver, writer

Community gardens can be large or small, independent or affiliated with an organization. If you can't find a community garden near you, why not start one yourself? First, determine what resources are available to you. If there's already a community or urban garden system in your city, you can utilize their expertise and start a garden with their assistance.

Community gardening organizations often provide assistance with the design and planning, garden construction, billing and maintenance of the utilities, and ongoing guidance for the life of the garden. The process of working with an organization to build a garden can sometimes take a while—as much as three years. They typically have a waiting list for new gardens, so it's important to act quickly if you'd like their help starting something in your neighborhood.

It's also possible to start a community garden on your own, without assistance from local community garden organizations. You may decide to create a garden for the benefit of your friends or co-workers, or you might want to align the garden with a group you're involved with. Neighborhood associations, religious organizations, gardening clubs, and Boy or Girl Scout troops are just a few examples of groups that might enjoy a community garden project. If the garden is meant to benefit a specific group, it's important to include them in all the planning right from the beginning. The ACGA has a set of tools and resources on its website for creating a community garden.

The Qualities of Good Land

Whether you're looking to start a community garden project or just want to create a garden for yourself, you're going to need access to land. Whether it's because of a lack of time, resources, or interest in keeping it maintained, weedy or bare land is a common sight in cities. You could find that well-maintained land can also be up for grabs. The property owner may be growing tired of investing the time and money in keeping the landscaping looking nice and might be interested in converting it into a garden if it means he or she can relinquish the responsibility of caring for the space.

Once you start looking for it, you'll see empty land everywhere, including the following:

- Front or backyards of houses and apartment complexes

- Land adjacent to churches, schools, or businesses

- Empty lots awaiting redevelopment

- Sections of city parks

- Land that's part of a large campus, such as hospital grounds or retirement communities

Any of these spaces can be converted into a garden. One of the great things about homesteading in the city is the abundance of land that's close and available—and often free!

However, just because you *can* garden in a space doesn't mean that you'll necessarily *want* to. Taking care in selecting your land will help prevent many potential problems.

Evaluating Potential Locations

You'll want to consider several factors when looking for land, including the following:

Proximity: A garden that's easy to get to is a garden that will be well cared for. A 20-minute drive across town might not seem like much right now, but if you multiply it by several trips a week, and throw in traffic, it may result in a neglected garden. Try to find a space that's close to where you live or along a route you travel often. If the garden is within walking or biking distance, that's even better!

URBAN INFO

You can get a bike trailer to haul your shovels and seeds behind you when you ride to the garden. Check out Chapter 20 for information on how to tow things with your bike!

Water: Be sure you have access to water for your garden space. Many empty lots don't have water spigots, so check this out before committing to the space. If the property does have water spigots, it's a good idea to test them to be sure they work properly. If you're planning on using a timer for watering, attach a hose cap to the spigot and turn it on to check for leaks.

Sunlight: As I discussed in Chapter 5's "Choosing a Location for Your Garden" section, unobstructed sunlight is critical to the success of a garden. Shadier gardens can be good places to grow leafy greens like lettuce or spinach, but you'll find that your tomatoes and eggplants won't develop as well as you'd like without that sunlight.

Landscaping: You'll need to consider the current landscaping in your potential garden space. Well-maintained sod presents a different set of challenges to a gardener compared to a weedy patch of wild land. Sod is difficult to remove, but you may not have many issues with weeds in that space (although you might have a problem with the grass return-ing). Weedy patches are relatively easy to clear and dig into, but you'll likely encounter a vigorous weed population for your first few years in that garden.

If trees are growing in your garden space, you can plant around them, provided you take care not to damage the roots. However, sometimes trees will have to be cut down, and the stumps will need to be removed or "killed" to prevent regrowth. Be aware, too, that a garden space near trees may contain hidden tree roots under the soil.

Killing a tree stump doesn't have to be as dramatic as it sounds. Start by drilling holes in the stump. Stuff the holes with a high-nitrogen fertilizer (like animal manure or fish meal) and keep the stump wet. The fungi in the stump will feed on the nitrogen, and the stump will decompose!

URBAN INFO

Cutting down a tree so that you can plant a garden may seem counterintuitive. However, urban homesteading is not about being "green" just for green's sake. It is about living sustainably within a city, which often involves producing more of what you consume closer to home. If you only have one piece of land available to you, it may be that a garden is a better use of that land than a tree (depending on the tree, of course). Ultimately, the choice of whether to try and plant around the tree, remove the tree, or pick a different spot for your garden is up to you.

Zoning Considerations

All the zoning issues I covered in Chapter 5, such as front-yard gardens and growing in right-of-ways, are also relevant when you're considering gardening on someone else's land. However, the bounty of potential growing spaces once you venture beyond your own property brings with it additional potential zoning issues.

ROAD BLOCK

Is the land you have your eye on located in a covenant-controlled community? Better check their regulations because they may have restrictions on back- or front-yard gardens above and beyond what's in your city's zoning code.

If you're considering land in a business district, there may be landscaping restrictions laid out by the city or by the local merchant's association. Begin by asking the business owner for the exact zoning designation for the property and then check your city's zoning code. You can also contact the president of the merchant's association to learn if there are any limitations on what can be done with the land you're planning on using. The odds are good that the merchant's association will love the idea of a garden in their area, but always check before you plant.

The opportunity to garden a huge yard space or an empty lot can seem like a tremendous windfall to an urban gardener. However, it might not be allowed under the local zoning code. Some cities have limitations on the amount of land that can be turned into a garden when it's part of a residential area. In addition, some cities don't allow agriculture as the primary use of a plot of land, which would restrict your ability to garden on an empty lot.

Determining If the Land Is Safe

Testing the soil before planting is a good idea no matter where you plan on starting a garden, but it's especially important when growing food on unfamiliar land. By-products from urban development, oil or fuel dumping, or the rupture of underground storage tanks are just a few of the ways in which soil can become contaminated. When you consider that the plants you're growing will be producing food for you (and others) to consume, taking steps to ensure the safety of the soil seems well worth the effort.

Begin by talking to the property owner about the history of the land. Include a discussion about the pesticides and chemical fertilizers that have been previously used on that space, if growing organically is important to you. Contact your local agriculture extension office, explain what you're doing, and ask which soil tests they'd recommend. If you have reason to suspect your soil may be contaminated, you could also contact your local Environmental Protection Agency office and request an Environmental Site Assessment.

Land that's contaminated is often referred to as a *brownfield*. The good news is that groups across the country are dedicated to studying safe ways to grow food in or on brownfields. If you learn that the land you were planning on using for a garden is contaminated, the simplest option may be to find another space for growing your vegetables. However, if you decide to build a garden on contaminated soil, you can work with your local agriculture extension to decide on safe growing practices for that space. I talk about some of the possible options in the "Growing Above Ground" section later in this chapter.

DEFINITION

A **brownfield** is land that's contaminated with low levels of hazardous waste or pollution.

Land-Share Agreements

You may begin your quest to find gardening space by selecting a piece of land you'd like to use and then contacting the property owner. Or you may advertise your services as a gardener to anyone with available land and then pick through what's offered until you find a space that meets your criteria. Either way, creating a good land-share agreement will be an important part of the process.

Finding Land

If it's your first time negotiating a land share, you may think it's going to be difficult to "convince" someone to let you use his or her land. While you might run into the occasional property owner who doesn't share your vision, it's likely you'll find people throwing land at your feet (metaphorically speaking) once they learn you're willing to turn it into a garden. Most land-share agreements involve giving some of the harvest to the property owner, which makes it even more enticing!

SMALL STEPS

Remember, the key to making gardening fun and manageable is to start small. Keeping your garden well maintained can be especially important when you're using someone else's land. An initial plot of 200 square feet can yield a lot of food, and you always have the option of expanding next year!

The simplest way to find land is to use your personal connections. E-mail everyone you know, let them know what you're looking for, and ask them to forward your e-mail to everyone *they* know. Check with your local sustainability groups to see if they've set up a land-share program. You can even put a post on Craigslist or Freecycle to make your wish for land known to the broader community.

If you've got your eye on a particular piece of land, make contact with the property owner and explain what you'd like to do. Supplying the person with a clear description of the planned garden, and photographs of similar urban gardens, would be helpful. If you don't know who owns the property, you can usually find that information in your city's property records database.

Creating a Land-Share Agreement

Taking the time to craft a clear land-share agreement helps things run smoothly and prevents problems down the road. While it might feel overly formal to draft a contract, the property owner will likely appreciate that you're taking the project seriously.

Possible elements of a land-share agreement include the following:

- Amount of space to be gardened
- What will happen to the existing landscaping
- Who is allowed access to the space
- Hours you're allowed to access the space
- Who will pay for the water
- Who will maintain the spigot/sprinkler system
- How the produce will be shared
- Use of fertilizers and pesticides
- Who will decide what will be planted
- Actions needed to keep pets out of the garden
- Responsibilities for fall cleanup

ROAD BLOCK

If you'd like to garden on a property that's occupied by a renter, be sure both the renter and the property owner sign the land-share agreement.

Let's take a look at a sample land-share agreement.

Sample Land-Share Agreement

The property at 123 Washington Street is owned by John Smith ("Owner"). The southwest corner of the property, with an approximate size of 300 square feet ("Garden Space"), at the above property will be landscaped and maintained by Jane Brown ("Gardener") from March to October of this year. The Owner understands that the Gardener will remove the existing landscaping within the designated Garden Space. This contract is only valid for this year and must be renewed annually. The Gardener is not responsible for restoring the Garden Space to its previous condition. Gardener will remove all garden plants at the end of the season.

The Owner is responsible for paying the water bill for this property. If the water bill for any month during the growing season is in excess of $30 more than the same month in the previous year, the Gardener will pay the additional amount due. Additional water usage must be due to gardening activities and not because of damage or malfunction to or within the property's water supply or other uses. The Owner agrees to maintain the water spigot in good working condition. If the water spigot breaks, the Owner will repair the spigot within 3 days so as to maintain healthy irrigation for the garden.

All vegetables, herbs, and edible flowers grown in the Garden Space will be harvested by the Gardener. Approximately 25 percent of the food harvested will be given to the Owner, on no less than a weekly basis. The Gardener is responsible for all garden planning and maintenance. The Gardener and her designated friends (Tom Williams and Maria Johnson, "Assistants") will be allowed access to the Garden Space between the hours of 8 A.M. and 7 P.M., Monday through Sunday.

The Gardener will consult the Owner regarding the types and varieties of vegetables that will be planted in the Garden Space. The Gardener will use only organic growing methods. No synthetic chemicals, pesticides, or fertilizers will be used at any time.

The Owner agrees to install a temporary fence or barrier adequate to keep the Owner's dog out of the Garden Space. The fence or barrier will have a gate or other opening, wide enough for a wheelbarrow, that allows the Gardener to enter the Garden Space.

The Owner shall neither hold, nor attempt to hold, the Gardener or her Assistants liable for any injury, damage, claims, or loss to person or property occasioned by any accident or condition to, upon, or about the property.

This is just an example, of course. You may decide to split your produce with the property owner 50-50, or you might find a property owner who is happy to pay the water bill without any limitations. The beauty of this kind of arrangement is that it is individualized, so don't feel that you have to follow one specific model.

Guerrilla Gardening

The term *guerrilla gardening* may inspire thoughts of black (or green?) suited commandos, swinging in on ropes bearing trowels and seeds. The reality is less dramatic, although not without its perils.

URBAN INFO

Johnny Appleseed is considered to be one of the earliest guerrilla gardeners.

Guerrilla gardening is growing on land that doesn't belong to you, without getting permission from the property owner. This can also include unauthorized growing on public land. Some guerrilla gardening occurs without notice, when especially enthusiastic gardeners run out of space and decide to quietly extend their planting beyond their own property. Other guerrilla projects are overt, with prominent public spaces suddenly transformed into gardens. The motivations for planting guerrilla gardens are as varied as the gardeners themselves, but the desire to "dig in" and create neighborhood change in a tangible way is often at the heart of the act.

Guerrilla gardening can occur in traffic medians, public right-of-ways, or back alleys, just to name a few places. The space may be publicly or privately owned, but guerrilla gardening often happens on land that's been abandoned or neglected.

Guerrilla gardening is not just limited to vegetables. It often includes flowers and greenery or hardy perennial plants that can survive without much attention. Some guerrilla gardeners even choose to plant trees! If you plan on doing some guerrilla gardening, it's a good idea to keep in mind which plants are (or have been) native to your area because they will likely thrive in your climate.

Guerrilla gardening poses unique challenges and some obvious risks. Gardening tasks often have to be done at night, and water must be carried to the garden. Property owners may suddenly decide to develop or landscape the space, leading to the destruction of your carefully tended vegetation. And last but not least, remember that planting on property that doesn't belong to you without permission is illegal. The consequences may not be severe, but it could still lead to an uncomfortable situation with law enforcement or security guards.

ROAD BLOCK

Vegetables that grow near areas with heavy traffic have a higher probability of contamination due to automobile fluid runoff and the absorption of toxic exhaust fumes.

Growing Above Ground

Great gardens don't necessarily have to have their roots in the ground. Some of the best urban farmers around have developed systems for building gardens on parking lots or concrete slabs. This takes more work up front in terms of procuring the materials for building the garden, but it can result in some spectacular transformations in an urban landscape. Having the skills to grow food on old playgrounds, basketball courts, parking lots, or concrete patios opens up a wealth of additional "land" in the city.

The first step in building a garden on concrete is to put down a layer of mulch 4 to 6 inches thick. Mulch made from old Christmas trees works wonderfully, and many cities offer free mulch giveaway programs in the spring. On top of the mulch, add about 24 inches of compost. This is a lot of compost, but it's necessary to give your plant roots space to grow. You can create the compost yourself (see Chapter 22), or you can purchase it in bulk from a landscaping supply company.

If you're planting on asphalt, you may need to take extra steps to protect your plants. Parking lots can be particularly risky because of the likelihood that oil and other chemicals have leaked out of cars and been absorbed by the asphalt. Tarps can be used to create a barrier between the ground and your garden, and placing cardboard below the first layer of mulch provides additional protection.

ROAD BLOCK

If you're planning a garden in a parking lot, be aware that zoning codes usually specify the amount of available parking spaces that are required for a business to operate. Even if the business is not currently open, a lack of parking may make it difficult to sell the building in the future.

Other techniques for growing gardens above ground include building raised beds (as discussed in Chapter 5). If you're creating your garden on a brownfield or a hard surface that may be contaminated, you can take the extra step of building your raised beds with a solid wood bottom and then setting the entire bed up on bricks. Alternately, the growbag gardening method often used for patio gardens (see Chapter 4) can be adapted for larger-scale growing on concrete or brownfields. You can purchase large growbags from a gardening supply company or make your own out of trash bags or burlap sacks.

Remember, if you're growing on a brownfield or land you suspect may be contaminated, you need to consult with your local agriculture extension office before planting—even if you're planning an aboveground garden.

The best way to keep the soil of your aboveground garden nourished is to put some worms in the beds. Red wigglers, the type used for worm composting, are the best ones to add.

When it comes time to plant, small plants like lettuce and spinach can be seeded directly into the compost. You may want to add an extra layer of mulch on top of the compost between large plants, such as tomatoes or squash.

The Least You Need to Know

- An abundance of land is available in the city for growing food.
- Community gardens provide opportunities to grow and learn alongside your neighbors.
- A carefully crafted land-share agreement helps ensure a successful partnership.
- It's possible to build a great garden just about anywhere, even on concrete or contaminated land.

Seed Starting in the City

In This Chapter

- Locating great places for growing seedlings
- Using natural and artificial light
- Raising seedlings, from sprout to garden
- Saving seeds for next year's crop

Great gardens consist of many things, but often vegetables like tomatoes, peppers, and eggplants are the main attraction. These plants must be first raised as seedlings before they can find a spot in your urban garden. You can find seedlings in several places, from garden centers, to local growers, to even Craigslist.

But seedlings can be expensive. It also might be difficult to find the specific varieties you're dreaming of growing. Hoping to try German Red Strawberry tomatoes? Moon and Stars watermelon? You're probably not going to find those anywhere except a seed catalog or seed exchange. Because of this, many gardeners begin the growing season well before the ground has thawed by starting their seedlings themselves.

In this chapter, I talk about finding space in (or outside) your home for growing seedlings. I discuss options for using both natural and artificial light, depending on which seedling setup best fits your situation. I review the steps for successful seed starting, all the way through transplanting the fully grown seedling into your garden. I conclude the chapter by covering the basics of seed saving, so next year you can start with seeds from your own plants.

Finding Places to Grow Seedlings

Sometimes, locating the best spot for growing seedlings can be almost as challenging as the actual growing itself. Some gardeners surrender to the notion that seedlings will take over their home for a couple of months in the spring, although that doesn't always have to be the case.

When you begin scouting locations for your seedlings, remember that their space needs increase drastically as they grow into maturity. A single seedling tray can hold 140 (or more) just-sprouted seedlings, but the same size tray will only support 15 2-month-old tomato plants. Of course, some of the seeds you start won't grow well, so it makes sense to plant more than you think you'll need. Just be aware that what begins small will inevitably get big, so plan accordingly!

Growing with Natural Light

Seedlings tend to be fussy about getting enough light. Without adequate light, seedlings will become "leggy," which means they grow a long, thin stem that's ultimately unable to support the plant.

You can grow seedlings indoors using window light if you'd like, following many of the same guidelines laid out in Chapter 4 regarding indoor container gardening. However, because seedlings are more fragile than mature plants, they need extra monitoring. Be especially careful not to put your seedlings near wall or vent heaters. Rotate seedling trays or pots a quarter turn each day so they don't end up permanently curving toward the light. South-facing windows are the best source for indoor sunlight.

SMALL STEPS

If you're growing just a few plants, you can likely find room for them on your windowsill. If your seedlings outgrow the windowsill, try placing a bookshelf near the window and using the shelves to hold your seed trays.

If you have access to outdoor space (like a balcony or a stoop) that gets at least six hours of sun a day, you can use that space to grow your seedlings—provided you give them some protection. Young seedlings aren't strong enough to handle the extremes of wind and unfiltered sunlight on their own.

The best way to provide seedlings the protection they need is to construct a little tent, like a mini-greenhouse, covered with clear plastic sheeting. The frame can be made of PVC pipe, wood, or anything else you have on hand. Just as when building a cold frame (see Chapter 9), it's important that the tent has the ability to be opened so the seedlings don't become too hot. (See Appendix B for a website that provides plans for building a structure like this.)

However, even the most well-constructed tent won't keep delicate seedlings warm enough at night. Once the sun begins to set, bring your seedlings indoors. Seedlings can take a small amount of evening cold, but not much. If you know you'll be away all day and into the night, it may be best not to put your seedlings outside that day.

Growing with Artificial Light

Growing under artificial light is the best way to ensure your seedlings receive adequate light. Artificial light is consistent, and you can control every aspect of it, as opposed to the variations in intensity and duration of natural light. Artificial light also allows those without outdoor space or good window lighting to grow seedlings. In addition, it can be a real space-saver. Individual seed trays can be tucked away in closets, and multiple seed trays can be positioned under lights on a wire storage rack or bookshelf.

It's critical that the lights be adjustable, since you'll need to raise the height of the lights as the seedlings grow. It's best to keep the lights 2 to 4 inches above the tops of the plants. The larger the plants get, the farther away the lights can be.

Lights should not be left on 24 hours a day. A cycle of 16 hours of light and 8 hours of darkness is best to simulate spring light conditions. If you want, you can purchase a timer from a hardware or lighting store so you don't need to control the lights manually.

Seed trays should still be rotated regularly because the light is sometimes not distributed evenly under the bulb.

Artificial lights can be a bit of an investment, but over time, they're still cheaper than purchasing seedlings from a garden center. Regular household light bulbs are not suitable for growing seedlings, in part because they emit too much heat. Florescent lights (sometimes called "shop lights") are an option some gardeners use, but they don't emit the full spectrum of colors plants need to flourish.

The best option is to purchase full-spectrum "grow lights," which are specifically designed for raising indoor seedlings and plants. Grow lights come in a variety of sizes and models and are available at garden centers, hydroponic suppliers, grow stores, and online. The number of lights you'll need to raise your seedlings depends on the number of plants you're growing, how large the lights are, and how your trays are arranged under the light(s). If in doubt about how many to purchase, call your local garden center and describe your setup. They should be able to advise you.

Basic Seed Starting

I have great faith in a seed. Convince me you have a seed there, and I am prepared to expect wonders.

—Henry David Thoreau

Space for raising seedlings can be hard to find, so it's good to know that every vegetable doesn't need to be started indoors. The garden planning chart in Appendix C includes information about which vegetables need to be started indoors and which ones can be sown directly outside, or directly into a large pot, if you're doing container gardening.

A good approach is to prioritize the vegetables you'd most like to grow and be sure you have enough space for those seedlings before adding anything else to your garden list. If you have limited space under your grow light but lots of window space, you can begin seedlings under the light and transfer them to the windows as they get bigger and are transplanted into larger containers.

Gathering Your Supplies

Successful seed starting is not all about the supplies, but having the right materials can make a big difference. Here's what you will need to get growing:

Containers: Seeds can be very close together when they're first planted. Some gardeners choose to use an open tray, called a "flat," to sprinkle their seeds. A typical flat is 21×11×3 inches. If you choose to make your own flats, be sure to create openings for drainage in the bottom and only use wood that hasn't been chemically treated. Another option for starting seeds is to find flats that have individual pockets, or "cells"—usually about 72 cells per flat.

SMALL STEPS

Landscaping companies, especially ones that install flower gardens, can be a great source for free containers. They purchase hundreds of pots of flowers at a time, and rarely do they have a use for the pots after they plant the flowers in the ground. Be sure to sterilize any used containers (including containers previously used by you) with bleach or vinegar water before planting.

But pretty much anything can be used for growing, provided it has holes for drainage. Reusing the bottom half of egg cartons, with small holes poked in the base of each divot, is a popular choice for starting seeds.

If you're acquiring all your containers at the same time, get some individual pots for when the seedlings outgrow their flats. Most plants can be moved into pots that are 2½ inches in diameter, but tomatoes can grow so large they eventually need 4½-inch pots.

Soil: Seeds grow best in a mixture that's light and airy. Commercially available seed-starting mixtures (which are soil-less) are a good choice. If growing organically is important to you, be sure the seed-starting mixture is labeled organic. When the seedlings grow big enough to need repotting, they can be planted in potting soil (also available organically). If you'd prefer to make your own mixture, use the DIY Potting Soil recipe in Chapter 4.

Fertilizer: Seedlings grow fast, and they need fertilizer. A good way to feed seedlings is to use a liquid fertilizer when watering. Choose one specifically for seedling growth. (Look for one labeled organic if that's important to you.) Hydroponic and grow stores usually sell the best liquid fertilizer, but you can also find it at garden centers. "Tea" made from compost or worm droppings can also be used for fertilizing seedlings. (See Chapter 22 for instructions for making compost or worm tea.)

Watering can: A simple watering can with a thin spout is a great tool for watering seedlings. The thin spout allows you to get under the plant leaves and water right at the soil line. Spraying water on the tops of the plants may be faster, but wet plant leaves can lead to disease.

Seeds: Seeds are available in all kinds of places—garden centers, hardware stores, grocery stores, and online. A typical seed packet will likely be more than you're wanting to plant—30 tomato seeds per packet, for example.

If you'd like to plant a variety of seeds but don't want to spend a lot of money, you can coordinate a seed swap with other gardeners in your area. Each gardener is responsible for purchasing just a couple of packets of seeds. Everyone shares and takes home a few seeds of each variety. If you do end up with more seeds than you can plant, they'll keep for a while, usually up to three years, if you store them in a cool, dark place. (I discuss saving seeds from your own plants, plus the differences between heirloom and hybrid seeds, later in this chapter.)

ROAD BLOCK

Even though seeds can last for three years, that doesn't mean you should purchase older seeds. Look on the seed packets for a "Packaged For" date of this year to be sure you're buying the freshest seeds available.

Heating pad: A heating pad under your seedling flat is an efficient way to raise the soil temperature. Some heating pads automatically shut themselves off after a couple of hours, so try to find one that will stay on consistently.

Soil thermometer: You can find these relatively inexpensively in a local garden center or online. A soil thermometer isn't required to start seedlings, but it can prevent you from over- or underheating your soil.

Plastic wrap: This helps maintain even soil moisture during the initial stage of planting.

Starting Your Seedlings

The garden planning chart in Appendix C helps you decide when to start seeds indoors, relative to your last frost date. (It's okay to begin your seedlings a bit later than the dates listed. Starting a garden late is better than not starting one at all!)

If you only have a small space available for seedlings, you may need to start all your seeds in the same flat. If this is the case, you can average out the various start dates and plant all of your seeds at the same time. This might mean some plants are a little larger and some are a little smaller when it's time to transplant into the garden, but it may be worth it if it means starting with one seed tray instead of several.

One important thing to remember throughout the various stages of seed starting is to carefully label your seedlings. It's not always easy to tell one plant from another without a label, and mixed-up seedlings lead to a mixed-up garden!

Seedlings do not need light until they *germinate*, or spout. However, they do need heat. Soil temperature should average around 77°F. Cooler soil will lead to lower rates of seed germination, and warmer soil could kill the seeds.

DEFINITION

Germination is the process by which a plant emerges from a seed and begins growth. If a seed is germinated out of soil, the first clearly visible sign of germination will be a little white sprout poking out of the seed. When seeds are germinated in soil, the appearance of green growth emerging from the soil demonstrates germination.

It's beneficial to do a "test run" of your seed starting setup to be sure the temperature is correct before adding seeds. Fill your seedling flat with seed starting mixture, and water the mixture until it's thoroughly wet but not flooded—it should have the dampness of a wrung-out sponge. Place the flat on the heating pad and turn on the power. Leave the tray

alone for a couple of hours and then return and test the temperature with the soil thermometer. Adjust the heating pad accordingly until the soil temperature is close to 77°F.

When you've got everything set up correctly, you're ready to begin! Here's what to do
next:

1. Rewater the seed starting mixture in the flat to achieve the desired wetness—
   remember, like a wrung-out sponge.

2. Place the seeds on the surface of the seed starting mixture. If your flat has
   individual cells, plant two seeds per cell. If you're using an open flat, plant seeds
   about an inch apart. (Seeds can be planted closer, but it makes repotting more
   difficult.) In general, plant two seeds for every one plant you hope to grow.

3. Sprinkle seed starting mixture over the top of the seeds. Seeds should be buried
   about as deep as they are long.

4. Lightly water again to wet the surface of the soil.

5. Cover the flat with plastic wrap and place on the heating pad. The plastic wrap
   should be loose, not tight, to allow a small amount of air circulation.

6. Monitor the soil dampness daily and water if necessary. While it's covered with
   plastic wrap, the soil should stay reasonably damp without extra watering.

7. Seed germination times vary depending on the vegetable, but most will show
   signs of growth within a week. When the seedlings begin sprouting, remove the
   plastic wrap. It's time to move the seedlings into the light!

The timing of the plastic wrap removal can be tricky because not everything will germinate at the same time. Sprouted seedlings can last for a day or two under plastic wrap
while you're waiting for the slower seeds to germinate, but then they begin to suffer. If
necessary, peel back the plastic from select parts of the flat, leaving other parts covered.
(Be aware, however, that the germinated seedlings won't grow very well until they're
exposed to light.) When you can't wait any longer, remove the plastic and move the flat
into the light, even if some things haven't sprouted yet. They still may germinate a day or
two later.

Maintenance and Repotting

Providing proper amounts of light and water are the two most critical elements of successful seed starting. I discussed guidelines for lighting earlier in this chapter, so now let's
talk water.

Plants should be watered only as needed—not on a fixed schedule. Check the soil moisture twice a day. It's okay for the top layer of the soil to dry out slightly between waterings, but if all the soil is allowed to dry, the seedlings will die. Use your watering can to water directly at the soil line, under the leaves of the plants.

Once the seeds have germinated, begin adding fertilizer to the watering can each time you water. If you're using commercial fertilizer, follow the instructions on the bottle. If you're using compost or worm tea, dilute it with 10 parts water to 1 part tea. The mixture should be a light amber color.

SMALL STEPS

Position an oscillating fan across the room from the seedlings so the air blows gently across them at regular intervals. The slight bending of their stems will help the seedlings become stronger as they grow.

The first leaves to emerge from a seed are called the *cotelydons*, or "seed leaves." The cotelydons don't look anything like the plant's mature leaves. As the seedling continues to grow, the "true leaves" will emerge. When the seedlings get large enough, they'll need to be repotted, or moved into bigger containers. The seedlings are ready for repotting when the leaves of neighboring seedlings touch. Select the largest and strongest seedlings. At this stage, it may be wise to repot a couple more seedlings than you'll need in the garden, just in case. Any seedlings that won't be repotted should be snipped at the soil line with scissors.

Containers 2½ inches in diameter are a good choice for repotting. Fill the containers with moist potting soil, and create a hole big enough in the center of the container for the roots of the seedling. Remove the seedling from the flat and gently place it in the new pot.

It's important to avoid handling seedlings by their stems. The stems are fragile, and if they're damaged at all, the seedling won't survive. Handle seedlings by the soil around their roots or by their leaves.

Plant the seedlings so the first set of true leaves are just above the soil level. This may mean that the cotelydons get buried, which is fine. Gently brush potting soil around the stem of the seedling and add a little fertilized water. A 2½-inch container is big enough for most seedlings until they're ready to go into the garden, but tomatoes may need to be repotted again into a 4½-inch container.

Troubleshooting

Seed starting can be challenging at first, and it takes most gardeners a while to feel they've mastered it. Here are some common problems, along with possible solutions:

Seeds don't germinate: Seeds that are too old, have become too hot, or got wet while in their packaging won't germinate. If you think this may be the case, see the "Saving Seeds in Urban Gardens" section later in this chapter for instructions on testing seed viability. If the seed starting mixture is too cold or too wet, the seeds will rot before they can germinate. If the seed starting mixture is too hot or allowed to dry out, the seeds will die. Dig up a seed and check to see if it's swollen and rotted, if it appears to have sprouted and died, or if it's sprouted and healthy—which means you just need to wait a little longer for the seedlings to appear.

Seedlings are leggy: This happens when seedlings aren't getting enough light. Adjust the position of the seedlings and/or the light as needed. It's possible to replant leggy seedlings deeper in a new seed tray, but it must be done carefully.

Seedlings topple over at the soil line: This is a condition known as "damping off," and it's the result of a fungus in the soil. It can also be brought on by soil that's too cold or too wet. Sterilize seedling flats and containers every spring before planting with a bleach or white vinegar water bath and never reuse seed starting mixture or potting soil. If the soil in your flat or container is contaminated, repot healthy seedlings immediately in sterile containers.

Seedling leaves yellowing: In this case, the seedlings are not getting enough nitrogen. Increase the amount of fertilizer added to the water.

Mold on the surface of the soil: This is an indication that the soil is too wet. Use an oscillating fan to increase air circulation and stop watering for a couple of days until the soil surface dries slightly.

Time to Transplant

The proper time to transplant seedlings into the garden, or into their final pot for container gardening, varies with each vegetable. For many of the seedlings started indoors (like tomatoes, eggplants, and peppers), it's around a week past the projected date of the last frost. These plants enjoy warmer weather and cannot survive frosts or temperatures that drop below freezing. (See the garden planning chart in Appendix C for the "earliest outside" date for each vegetable.)

Even if it's plenty warm outside, moving a seedling directly from its sheltered place indoors into the garden will almost certainly result in the death of the plant. This is a condition known as *transplant shock*, and it's caused literally by the shock of moving too quickly from inside to outside.

The key to avoiding transplant shock is to *harden off* your seedlings before transferring them into the garden. Hardening off involves gradually exposing your seedlings to the temperature variables, wind, and unfiltered sunlight of the outdoors. This process should take about a week.

DEFINITION

Hardening off is the process of gradually exposing seedlings to outdoor conditions before planting them in a garden or outdoor container.

On the first day, the seedlings should be placed in the shade, at least partially sheltered from the wind, for a couple of hours. Over the next few days, the amount of time the seedlings are outside can be extended, and they can spend greater portions of their outdoor time in direct sunlight. Near the end of the week, the seedlings should be outdoors for a full day, and they should spend a full 24 hours outside before being transplanted into the garden.

When the plants have been hardened off, they're ready to go into the garden. The procedure is identical for most seedlings (the exception being tomatoes):

1. Dig a hole slightly larger and deeper than the pot the seedling is growing in.

2. Put a handful of compost in the bottom of the hole.

3. Gently remove the seedling from the container, and place it in the hole. The seedling should be planted deeply enough that the first set of true leave are just above the soil level.

4. Pour water on the plant roots.

5. Gently fill the hole, making sure the seedling remains upright.

Now, what about those tomatoes? Instead of digging a hole for the seedling, dig a trench that's almost as long as the seedling is tall. Add compost and lay the seedling down in the trench, gently curving the plant up so just the top crown of leaves is above the soil level. Pour water in the trench and fill it with dirt. The stem of the tomato plant will form new roots when it's buried in the soil, so your seedling will develop a strong root structure to support the growth of the plant.

Saving Seeds in Urban Gardens

Seed saving is the final step in the cycle that connects one season of gardening to another. When you save seeds, you naturally harvest from the healthiest and best plants you have. Therefore, over time, you'll end up with strains of vegetables accustomed to your climate and soil. Seeds are a renewable resource, and saving your own can make gardening more affordable.

Also, the selection of seeds available commercially isn't that large, especially compared to the incredible number of vegetable varieties that used to exist. Many gardeners see seed saving—and sharing—as a way to help preserve biodiversity.

Heirlooms and Hybrids

Heirloom (also called open-pollinated) varieties are plants that maintain their traits from one generation to the next. As long as they don't accidentally cross with another variety during pollination, the parent plant will produce offspring (seeds) that are identical to itself. Heirloom varieties are never genetically modified. Heirloom vegetables are available in a wide variety of shapes, colors, and sizes, and many gardeners feel that heirlooms also offer the best varieties of flavors.

Hybrid plants are the result of cross-breeding between two compatible varieties to create a plant with specific characteristics. Some hybrid vegetables are bred to produce earlier, grow larger, or keep well so they can be shipped over long distances. Because hybrid varieties are the result of a cross, their offspring do not retain the characteristics of the parents.

 DEFINITION

Heirloom varieties are open-pollinated and maintain their traits from one generation to another. **Hybrid** varieties are the result of a cross between two similar plants and do not produce seeds that maintain the characteristics of the parent plant.

Hybrid vegetables don't reliably pass along their traits through their seeds, so they're not recommended for seed saving. It's not that the second-generation plants won't grow, but you can't be sure of the quality of what they'll produce. Heirloom varieties are the best to grow if you intend to save seeds.

There's an extra challenge with seed saving for urban gardeners. As I mentioned earlier, heirloom varieties will produce true to type *if* they don't accidentally cross-pollinate with a similar vegetable nearby. Rural seed savers intentionally isolate each variety they grow, leaving a designated distance between varieties to ensure purity. Sometimes the necessary distance is 50 feet, and sometimes it's 500 feet. At any rate, this is not practical for an urban gardener, who likely wants to grow several varieties of different vegetables in a small space.

Starting with heirloom seeds gives you the best chance for good seed saving. If there is some cross-pollination over the course of the summer, you may end up with some unexpected vegetable varieties when you plant the following spring. Who knows—you may end up with something you really liked! If not, you can just acquire some new seeds and start over with a fresh seed line.

How to Save Seeds

The techniques for saving seeds depend on the plant and how it produces its seed. Certain types are easy to dry and save—in fact, the plant will do it for you if you leave it alone long enough—and others are more challenging. Always store saved seeds in a cool, dry, dark place.

If you're unsure whether your saved seeds are still good to plant, there's a simple way to check seed viability. Roll a few seeds in a damp paper towel and place the towel in an open plastic sandwich bag. This keeps the towel moist but helps prevent it from molding. Keep the seeds at room temperature and in a dark place (sticking them in a kitchen cabinet is a good option) for a few days. Unroll the paper towel and check to see if the seeds have germinated. You should see a small white sprout emerging from the viable seeds.

Here are the basic techniques for some of the most commonly saved seeds:

Beans and peas: These are great for beginning seed savers because the seeds are easy to find—they're the actual beans and peas you would eat. Instead of harvesting, let the pod stay on the vine until it dries. Open the pod and remove the dried seeds. If frost is coming before the seeds have a chance to fully dry on the vine, pull out the entire plant by the root and hang it upside down inside to finish drying.

Cole crops—broccoli, kale, and arugula: Allow the plants to "bolt" (send up a long stem) and flower. Eventually, multiple small seed pods will form along the flower stem. The pods need to dry out fully for the seeds to be viable. However, the pods are fragile, so don't leave dried pods alone for too long or the seeds may scatter in your garden. Pick off pods individually, or pull the entire plant. You can place the pods in a bag and smash them to get the seeds, or you can hold the entire plant inside a large bag and shake it until the seeds fall out of the pods. The dried pod casings can remain mixed in with the seeds; they won't cause any problems.

Lettuce: The plant will bolt and create flowers. Wait until half of the flowers have dried out and gone to seed and cut off the top of the plant and allow it to finish drying in an open paper bag. Or you can harvest small amounts of seeds periodically from a bolted lettuce plant by pinching them off the dried flower heads.

Squash: The squash must be fully mature before the seeds can be harvested, so even summer squash needs to stay on the vine until the outer skin has become hard. In general, leave the fruit on the vine for three or four weeks past the intended harvest date to give the seeds time to fully ripen. When it's ready, cut open the squash and scoop out the seeds. Rinse the seeds thoroughly in a wire strainer, removing all the squash flesh. Dry the surface of the seeds with a towel and spread them on a tray until they fully dry.

SMALL STEPS

Finding space to dry seeds can be a challenge. Stackable letter trays (the kind sold at office supply stores) can allow you to dry multiple batches of seeds without taking up a lot of space.

Peppers: Allow the peppers to mature fully. Most peppers (even types that are typically eaten green, like Anaheims) turn red when fully mature. Cut the pepper in half, and gently scoop out the seeds. Spread the seeds on a tray, and allow them to dry.

Tomatoes: Allow the tomato to ripen fully. Cut the tomato in half horizontally (at the equator), and squeeze the jellylike substance that contains the seeds into a jar. Add a little water, cover the jar, and set aside at room temperature in a dark place for about three days. Shake the jar gently once a day. A layer of fungus will appear on the top of the mixture, but don't be alarmed. The fungus helps protect the seeds from diseases. After three days, fill the jar with water. The viable seeds will sink to the bottom. Pour off the water, and everything floating on top of the water, and strain out the viable seeds. After the excess water is completely strained out, spread the seeds onto newspaper or paper towels and allow them to dry fully.

The Least You Need to Know

- Raising your own seedlings can help save money and gives you a way to grow unique vegetable varieties.
- Adequate lighting is crucial for seedlings, but they're not picky about whether that lighting is natural or artificial.

- Water and heat must be monitored to keep seedlings healthy—too much or too little of either will cause them to fail.
- Seed saving is a way to make your garden self-sustaining from year to year, but urban gardeners may unintentionally create new varieties due to cross-pollination.

Keeping Your Garden Healthy

In This Chapter

- Simple techniques for weed prevention
- Making the most of your weeding time
- Outwitting garden pests and critters
- Preventing and dealing with plant diseases
- Diagnosing nutrient deficiencies

Part of the fun of growing a garden is envisioning healthy plants that are thriving and producing tons of beautiful food. It's not nearly as enjoyable to focus on the nasty things that might harm your garden. Nonetheless, these challenges are a part of the process of raising food. Armed with knowledge about how to deal with weeds, diseases, pests, and soil deficiencies, you can approach the growing season prepared to give your plants the support they need to flourish.

In this chapter, I begin by examining ways you can prevent weeds in your garden, because this is truly an area in which an ounce of prevention is worth a pound of cure. To deal with those hardy weeds that do manage to take root, I discuss how to prioritize your weeding so you can use your time effectively. I talk about the garden pests that might invade your garden and the sometimes cute (but usually not) critters that may also appear. Finally, I review strategies for preventing and dealing with plant diseases, plus how you can diagnose nutrient deficiencies that may be harming your garden.

Preventing Weeds

Strictly speaking, a weed is any plant that's growing where you don't want it to. Weeds are problematic because they compete with crops for moisture, sunlight, nutrients, and growing space. The extent to which you will likely battle weeds depends on the area where you've put your garden. Whatever was growing in the ground before the garden was installed will probably try to return.

Raised beds offer a head start against weeds because they're typically filled with fresh soil that's weed free. However, because weed seeds can be spread by many things—such as wind, water, animals, soil amendments, and even garden tools—they're sure to make an appearance in every garden. Perfect weed prevention is impossible, of course, but the best way to deal with weeds is to try to stop them from growing in the first place.

URBAN INFO

Dandelions and thistles have earned reputations as challenging weeds, but the most difficult and pervasive weed in many urban gardens is a plant that others carefully cultivate: grass.

Mulching

Weeds, like all plants, need light and air to grow. If they're effectively smothered, they'll die or fail to sprout in the first place. That's where mulch comes in. A mulch is a covering placed on the surface of the soil with the goal of preventing weeds. Mulches can help conserve moisture and regulate soil temperature, too.

Any number of substances can be used as mulch, and some are readily available—and often free—to urban homesteaders, including the following:

- Grass clippings

- Old leaves

- Cardboard

- Newspaper

Cardboard and newspaper should be watered after being placed on the soil, held down with rocks or boards at the edges, and covered with straw. You may notice that a couple of these items—grass clippings and old leaves—were also mentioned in Chapter 5 as possible soil amendments. The difference is in the application. Small amounts that are dug into the soil help to amend it, but large amounts piled on the surface of the soil can act as mulch.

In addition, garden centers sell landscaping fabric or plastic sheeting, which can also be used as mulch.

ROAD BLOCK

Whenever purchasing straw for use in or near your garden, be sure it's classified as "weed free." Otherwise, the straw may bring weed seeds into your garden, multiplying weed problems instead of reducing them!

Bare dirt is a magnet for weeds. Mulch can be used to help block weeds and retain moisture wherever there's open space in your garden. Mulching is also one option for preventing weeds in garden paths. (Another possibility is cover cropping.) In addition, you can use mulch in the garden beds between the plants. Leafy vegetables planted close together usually don't need to be mulched. However, larger plants like tomatoes or squash that are spaced farther apart so they have adequate room to grow leave lots of open dirt when the plants are young and small. Mulching between just-planted seedlings can help keep down weeds while the garden is taking root.

Cover Cropping

I briefly discussed cover cropping in Chapter 5 when I suggested planting Dutch white clover in garden paths as a way of adding nutrients to the soil. Cover crops have the additional benefit of acting as living mulches. They generally germinate quickly, preventing weeds from taking root and growing. Farmers use either legumes (clover, alfalfa, soybeans) or grasses (buckwheat, ryegrass, oats) as cover crops. For the purpose of mulching a growing garden, clover is probably the best choice because it's low growing. Clover forms a thick root structure that effectively blocks weeds, and its flowers attract bees and other beneficial insects.

Clover can be such an effective mulch that it shouldn't be used in every situation. Garden paths are a great place to cover crop with clover, and it can also be planted between large vegetable plants. However, smaller vegetables can be swallowed by aggressive clover, so don't plant it with your lettuces.

A good time to cover crop your entire garden is in the fall after the harvest. A combination of a legume and a grass gives your soil a good balance of nutrients while protecting the soil from weeds. In the fall, mow down the cover crop and either till or dig it into the soil. This can also be done in the spring, but if you wait until the cover crop has come back to life, it will be more difficult to turn.

Using Heat

Gardeners are always advised to store their seeds in a cool place because exposure to high temperatures will keep the seeds from sprouting. The same holds true for weed seeds, so one way to stop weeds from growing is to turn up the heat—a process called *solarization*.

DEFINITION

Solarization is the process of using the heat of the sun to raise soil temperature to a point that kills weed seeds, as well as soil-dwelling pests and diseases. This is accomplished by securing plastic over the surface of the soil for a sustained period of time.

Solarization involves securing heavy-duty plastic over the top of the soil, which amplifies the heat of the sun and raises the soil temperature significantly. This kills not only weeds and weed seeds but soil-dwelling pests and diseases as well. The process takes around six weeks if done during the summer. Solarization works best during the hottest part of the year, but it can also be done during cooler months if the plastic is left on for an additional couple weeks. Clear plastic is the best choice, and it's important that it be tightly secured with rocks or boards so heat can't escape.

Solarization is not the only way to utilize heat for weed control. Pouring boiling water over weeds kills both the plant and any seeds it's carrying, but unfortunately, it will likely kill everything else around the weed as well. Boiling water is a good solution for killing the pesky weeds that grow in sidewalk cracks, though.

One of the newer weeding tools is a propane-fueled torch, called a flamer. It literally cooks the weeds and is very effective. A flamer would work for eliminating weeds before planting but not for spot weeding because it would also damage anything growing near the weeds. Flamers present obvious safety concerns, especially if used in an urban setting.

A Busy Person's Guide to Weeding

Perhaps in a perfect world, everyone would have the time to keep their garden completely free of weeds. Other obligations often intrude, however, so it's important for gardeners to know how to prioritize and make the most of their weeding time.

If you only have a few minutes in the garden, your first task should be to remove any weeds that are flowering because that means they're about to produce a whole new crop of weed seeds. Next look for weeds that are actively encroaching on crops—especially big weeds that are shading small vegetable seedlings, or viny weeds (like bindweed) that like to strangle other plants. If you have more time to devote to your garden, you can remove all the other weeds that might be present in between crops, in the paths, or along the garden perimeter.

The best way to remove weeds is to dig them out, trying to get as much of the root as possible. Pulling the top of the weed, leaving the root in the ground, is a temporary solution, but you can bet the farm the weed will return. Any number of tools are available for removing weeds. The best ones help you dig out the weed while disturbing as little soil as possible, so the long and skinny "dandelion diggers" are often a good choice. If you're trying to cover a lot of space in a little time, you can use a hoe to chop the weeds down.

Once you've removed the weeds, you still need to handle them carefully so they don't spread any seeds they may be carrying. Unless you're an ace composter, it's not advisable to try to compost weeds. Your compost will be full of weed seeds unless you get the pile hot enough to kill them off. (See Chapter 22 for more information about composting.) If you have backyard livestock, they'll be happy to eat the weeds, and they'll even "compost" them for you in just a day or so!

Above all else, don't get discouraged. Weeds are a fact of life for every gardener, and their presence doesn't mean you've done something wrong. Weeds may be active in the spring and early summer, but they don't grow as well once the weather gets hot. A new garden may have lots of weeds during its first year or two, but if it's a well-tended garden, the weed problems will decrease somewhat over time. Fewer weeds equals fewer weed seeds, which means even fewer weeds the next year.

You can even turn your problem into an opportunity by learning which of your weeds are edible. People pay a lot of money for organic dandelion leaves at upscale grocery stores, and you might have a great crop of them growing right in your garden! See Chapter 23 to learn about edible weeds.

Managing Pests

Insects crawling all over your beautiful tomato plants may seem like a catastrophe, but bugs are not always a negative thing in a garden. Sure, some pests would like to munch your lettuce leaves, but beneficial insects help keep the bad ones in check. For this reason, annihilating every insect through insecticide spraying or other drastic means won't be good for the overall health of your garden.

The best defense against destructive pests is to keep your plants as healthy as possible. Plant in well-amended soil and be sure the vegetables get enough light and water, and they'll have a better chance of standing up to bugs. Companion planting (see Chapter 5) can be a great way to deter pests. A long row of carrots looks like a buffet to carrot-loving pests, but interplanting carrots with onions will confuse them and lead to less vegetable damage. Also, rotating crops each year is helpful because bugs leave behind eggs near their favorite plants. If you replant the same vegetable in the same spot, the baby bugs will have an easy time finding their food source.

SMALL STEPS

Diseased or pest-ridden plant parts should never be composted. Throw them away. This prevents them from reinfecting the garden. All is not wasted, though; chickens love some pests, like aphids, caterpillars, and slugs!

The incidences of different types of pests varies depending on where you live, so you can contact your local agriculture extension if you want to learn more about which bugs are prevalent in your area. In fact, pest outbreaks can be neighborhood specific, so it's also a good idea to make friends with other gardeners who live near you so you can share information with them.

A few pests seem to pop up frequently in urban gardens, but luckily, there are often simple ways to deal with them:

Aphids: These tiny little insects, which are only about ⅛ inch wide, are famous for attacking crops in the Cole family (like kale and broccoli), but they can actually feed on any number of plants. Aphids often clump together on the underside of leaves, and curling leaves can be a sign of aphid infestation. Spraying soapy water on aphids will suffocate them. However, it can suffocate beneficial insects, too, so be careful where you spray.

Speaking of beneficial insects, ladybugs love to eat aphids. You can purchase a bag of ladybugs from your local garden center and release them in your garden to help keep down aphids.

Ultimately, the best way to control aphids is to monitor your plants and remove any leaves or stems on which you see aphids. They spread quickly, so one aphid-infested kale leaf can take down your whole kale bed if it goes unnoticed.

URBAN INFO

Ants like aphids because they emit a sweet substance that ants enjoy. In fact, aphids are sometimes compared to cattle, with ants playing the role of rancher and protector. If you see a lot of ants in your garden, it's a sign you might also have aphids. Discouraging the ants (for example, flooding their nest with water from a hose) helps with aphid management.

Caterpillars and beetles: These take many forms, including cabbage loopers, Colorado potato beetles, and tomato hornworms. The simplest way to remove these—and any other large-ish bug—is to pick them off your garden plants. Carry a bucket of soapy water with you into the garden and drop any unlucky pests you find and pick off into the bucket.

You could also dust your plants with a bacterium known as Bt (*Bacillus thuringiensis*), which kills caterpillars and beetles by rupturing their stomachs. It's safe for humans, but it should be used with caution because it will also kill butterfly larvae.

Snails and slugs: These pets might be slow, but they are destructive. Snails and slugs feed on the leaves and fruits of plants, especially during wet weather or just after the garden has been watered.

An easy way to deal with these pests is to place shallow pans of beer in the soil near the affected plants. The snails and slugs will climb into the container and drown. You may endure teasing from your neighbor about the beer fest going on in your garden, but it's well worth it to be rid of the slugs!

Flea beetles: These little guys love to munch tiny holes in vegetable plants. Floating row covers, lightweight fabrics you can secure over the tops of plants, keep out pests like flea beetles but allow sunlight and water in. The plants must be covered early in the season, though, or the bugs will have already set up shop in the garden.

City Critters

It's not uncommon for urban gardeners to look longingly at the expanses of land available outside the city and covet those open spaces away from noise and traffic. It's true that rural gardeners do have more room for planting, but urban folks have them beat when it comes to the problem of critters eating the vegetables. Gardeners outside large cities often have to deal with deer plus large numbers of crafty mammals and birds. That being said, there is no shortage of urban animals that enjoy eating gardens. A certain amount of vegetable theft is inevitable, but you can take steps to help protect your plants from specific critters.

Squirrels: By far the most ubiquitous of the city animal pests, squirrels are clever and persistent. They especially love corn but will also eat away at your squashes and tomatoes. The first step in preventing squirrels is to ensure you're not leaving extra food lying around for them. Bring pet food inside and be sure no fruits or nuts from surrounding trees are left on the ground. Squirrels don't enjoy cayenne or garlic, so mixing some (or both) into water and spraying it on your vegetables can help deter them. Plants grow quickly, so you should reapply the spray often.

Squirrels are impressive climbers and can scale fences, but a garden fence with electric wire at the top can be a good deterrent. Some people go so far as to purchase a motion-detection sprinkler for the garden because a sudden burst of water will frighten squirrels away. Many garden centers also sell small squirrel traps that can be used for capturing and relocating the animals, which many gardeners believe is the only real solution for these animals.

ROAD BLOCK

Before trapping any animal, it's wise to check with your city's Animal Control department to be sure there are no restrictions on wildlife relocation. Plus, they can likely give you tips on safe handling to prevent you (and the animal) from getting hurt.

Birds: Birds are most problematic for gardeners who grow fruit trees, grapes, and berries, but they can also pick away at tomato and corn plants from time to time. The most effective way to keep them out of fruit trees is to purchase bird netting and keep the tree covered from the time the fruit first appears until it's ready to harvest. Grapes and berries can also be wrapped in bird netting. Vegetable gardens can be draped with the netting or floating row covers, but it may be more practical to only cover the specific plants the birds are targeting.

Birds startle easily, so some gardeners suspend items that flash, like CDs or aluminum pie tins, from strings and hang them around the garden. You could always construct an old-fashioned scarecrow, which would be a fun addition to your garden, even if you didn't have birds!

Rabbits: Rabbits can be kept out of a garden by a well-made fence. However, because they like to burrow, the fencing must be buried at least 6 inches underground. Rabbits can also be deterred by human or dog hair—just visit the salon or the dog groomer every few weeks and ask them for hair. Sprinkle it along the perimeter of the garden to discourage the rabbits from entering.

Some gardeners also recommend planting clover around the boundary of your garden because rabbits love clover and will eat it instead of your vegetables.

Groundhogs: These can also be kept out by a fence, but it will need to be at least 4 feet tall and buried 18 inches into the ground. If possible, bend the top part of the fence outward to keep the groundhogs from climbing over it. In addition, many of the same strategies that work for squirrels (electric fencing, cayenne/garlic spray, and trapping) also work for groundhogs.

Mice: These rodents are attracted by easy sources of food, just like squirrels, so be sure to keep the area around the garden clean. If they become a problem, the easiest solution is an old-fashioned mousetrap. Be careful to leave it in a place where you (or anyone else rooting around in your garden) won't accidentally activate it.

Plant Diseases

Most plant diseases are difficult—and sometimes impossible—to cure, so it's better to focus your attention on preventing them from occurring. Once again, as with managing pests, keeping your plants strong is the best defense against diseases. Other possible preventative steps include selecting plants bred to be disease-resistant—these are usually hybrid varieties.

A number of diseases grow and are spread best on wet plants, so you can reduce your chances of sick plants by watering with a drip irrigation system that waters only the soil, not the plants. If you're using a sprinkler system, try watering in the early morning so the plants don't sit all evening with wet leaves. However, morning watering does lead to more water evaporation from the soil compared to evening watering, so it's a bit of a toss-up. Finally, you'll be more likely to catch any disease problems before they spread if you regularly spend time in your garden and inspect the plants carefully.

If plant diseases do strike, remove the affected parts and throw them away. If the entire plant looks sick, dig it and pull the whole thing—roots and all—out of the garden. If you're not sure what made your plant sick, save one of the infected leaves. Do a little investigating online or ask a gardening friend, and it's likely you'll be able to discover the culprit.

Here are some of the common diseases that strike urban gardens:

Leaf blight/spots: This appears as brown spots on leaves that grow in size until the leaf dies and falls off. It's caused by a fungus that spreads in wet weather or in gardens that are watered from above, so watering in the morning can be helpful. The fungus can live through the soil over the winter, so removing old plants in the fall and rotating crops in the garden is advisable. Organic copper-based fungicide sprays can be used if necessary.

Fusarium wilt: This fungus causes plant leaves to yellow and wilt, and then the plant may die. Once a plant is infected, there's no cure, but building healthy soils (see Chapter 5) reduces the likelihood of infection.

Powdery mildew: A plant disease that looks just like it sounds, powdery mildew appears as a fine white coating on plant leaves. It weakens the plant and inhibits its growth. Infected leaves will turn brown and fall off, and infected plant blossoms won't produce fruit. Powdery mildew can attack many varieties of plants, but in vegetable gardens, it shows up most frequently on members of the squash family. Spraying plants weekly with either milk or a baking soda solution can be effective in preventing the problem.

Insufficient Nutrients

Unhappy plants are not always the victims of pests or disease. Sometimes the culprit is a lack of adequate nutrients in the soil. If you take steps to build healthy soil, your plants should have everything they need to thrive.

If the soil is lacking, the plants will let you know. A nitrogen deficiency causes stunted growth and a yellowing of plant leaves. Too little phosphorus leads to a red or purplish color on the underside of leaves, along with stunted growth and delayed maturity. A lack of potassium causes the edges of plant leaves to turn brown and die, and the plant will have a tendency to wilt readily. A calcium deficiency can cause blossom-end rot, in which the bottom end of the plant's fruit develops a lesion.

ROAD BLOCK

Your front-yard garden may be exposed to an additional nitrogen source you hadn't planned on. If your plants are in an area that gets a lot of dog-walking traffic, be on the lookout for damage caused by nitrogen "burn" from dog urine, which will cause plants to turn brown and could possibly kill them.

Plant difficulties due to insufficient nutrients serve as a warning to let you know there are problems with your soil. If you're uncertain or would like more information, you can always get a soil test through your local agriculture extension office.

If you take steps to amend your soil to correct the nutrient imbalance, it will help prevent future plant problems and may give the current crops what they need to bounce back. See "Creating Healthy Soil" in Chapter 5 for recommendations on specific soil amendments for different nutrient deficiencies.

The Least You Need to Know

- Weed problems can be minimized by covering or planting any open spaces in the garden.
- The best defense against diseases and pest infestation is to give plants the water, light, and nutrients they need to grow strong and healthy.
- Many small animals will want to share in your harvest, but if you create protection for your garden, you can minimize the damage.
- Plants will show signs if the soil is lacking specific nutrients, and soil amendments will be needed to correct the imbalance.

Enjoying Your Bounty

Chapter 9

In This Chapter

- Garden vegetable harvesting tips
- Preparing for the first frost
- Extending your season with cold frames
- Donating or selling your extra produce

Food production is a core part of urban homesteading. Eating is an activity we engage in three times a day (at least), so attending to the source of one's food goes right to the heart of self-sufficiency. Reaping your harvest is—like every other part of growing food—not without its tricks. It's also possible to continue your home food production long past the end of summer and use some of your excess harvest to support yourself and your community.

In this chapter, I go over the basic harvesting techniques for many of the vegetables you're likely to be growing in your garden, plus I offer tips on the edible plant parts you may have otherwise discarded. I explore methods for growing food in cold weather and let you know which crops continue to ripen even if they're harvested prematurely. As an efficient urban gardener, you might be growing more than you can eat, so I also talk about how you can donate, trade, or sell the extras from your garden.

All About Harvesting

After all the time spent planning the garden, preparing the containers or beds, starting the seedlings, planting, tending, and weeding, the act of harvesting is especially rewarding. It may seem like a simple thing to pluck a tomato or pick a salad, but the way in which you do it makes a difference in the quality of your harvest and the overall longevity of the plant.

Timing is important, too, so in the following sections, I talk about the best time, within the development of each type of plant, to begin harvesting. It's important to give seedlings enough time to mature before you begin picking from them. However, if certain plants are left in the ground too long without trimming, they'll go to seed and become inedible.

Even the time of day you harvest can make a surprising amount of difference. Strolling out in the garden with your basket and greens knife to pick lettuce at 2 in the afternoon may not seem like it would be a problem, but your plants won't enjoy it and your salad might be lackluster. Heat is stressful for plants. You can see leafy greens and big squash leaves droop before your eyes during the hottest part of the day. Cutting off part of a plant when it's stressed is adding insult to injury. Plus, the greens in your harvesting basket will wilt more quickly if they're picked when it's hot. The best time to harvest is in the morning or evening when it's cooler. Mornings can be especially advantageous because that's when your vegetables will have the highest water content.

 ROAD BLOCK

There's one important exception to the "harvest in the morning" rule. It's not a good idea to harvest (or weed, for that matter) in the garden right after it's been watered. As discussed in Chapter 8, disease spreads faster between plants when the leaves are wet.

Having the right tools on hand makes harvesting easier. A sharp knife, preferably serrated, is useful when gathering certain spring vegetables and summer squashes. Inexpensive steak knives typically work well. A pair of scissors comes in handy when trimming herbs. A garden trowel is sometimes helpful when trying to harvest stubborn carrots. Any number of tools can be used to dig potatoes, including shovels, trowels, and pitchforks. If you're so inclined, you can invest in a large tool called a broadfork, which not only unearths potatoes but can be used for digging garden beds in the spring, too.

Harvesting Spring Greens

These vegetables are the joy of homesteaders because they're among the first homegrown vegetables to make an appearance on your dinner plate. We tend to think of lettuce harvesting as a one-shot deal, compared to ever-giving tomato branches or squash vines, but these spring vegetables can continue to flourish and provide food for many weeks. Often it's the heat of the summer or an infestation of pests that causes the plants to stop producing rather than them deciding on their own that they're done.

Salad greens: This bountiful category of garden vegetables includes lettuces, spinach, arugula, mizuna, endive, frisée, and radicchio. The leaves are ready to be harvested as soon as you determine that they're salad sized.

Harvesting can be done in one of two ways. You can gently remove individual leaves from the plant. The outer leaves are more mature, and the inner leaves are still forming. Picking off outer leaves allows the plant to continuously regenerate itself.

If you'd like to take a larger quantity of leaves at once, use a knife to cut off the entire plant at its base, about an inch or so above the soil. This leaves the bottom part of the plant—usually a stub with a couple of leaves or stems attached—in the ground. If you leave the stub undisturbed, you will most likely enjoy a regrowth of the plant.

Sometimes gardeners choose to remove the entire plant, roots and all, as a way of *thinning* a particularly dense garden bed to achieve the desired plant spacing. If salad greens are left too long without being trimmed, they will *bolt*, or send up a flower stalk, and start making seeds. Almost all salad greens become extremely bitter and inedible once they've bolted, and the only options are to remove them and replant or allow the seed development for seed saving. The exception to this is arugula, which is still edible, but considerably spicier, after the plant has bolted.

DEFINITION

Thinning is the practice of selectively pulling plants out of a crowded garden bed to give the remaining plants more room to grow. The term also applies to removing some of the fruit from a fruit tree to support the health of the tree and the growth of the remaining fruit. **Bolting** occurs when a vegetable plant sends up a flower stem for the purpose of creating seeds. Plants halt their regular growth after they've bolted.

Cooking greens: This delightfully healthy bunch includes kale, chard, collards, mustard greens, and bok choy. Just like salad greens, the leaves on the outer part of the plant are the most mature.

Gently snapping off individual leaves is the best way to harvest from these plants. Start at the outer edge and work inward, taking as many leaves as you need. It's best to leave at least a few baby leaves in the center of the plant so it has a head start on regrowth. It would likely grow again even if you were to take a knife and cut off everything but the stump, but it could take a while to return to an edible size. Taking only the most mature leaves and leaving the rest behind provides you with a steady supply of greens.

Cooking greens allowed to grow too long without trimming will bolt. The entire plant can be cut back if this happens, and it will usually regrow new, smaller leaves.

Broccoli, cauliflower, cabbage: These members of the Cole family grow as a central head surrounded by a ring of leaves. The size of the mature head depends on the variety, so it's good to know ahead of time whether you can expect your broccoli to grow 3 inches or 6 inches in diameter before harvesting.

Use a knife to trim the head at its base. If you leave the rest of the plant in the ground, it will typically sprout one round of smaller heads off the main stem. The leaves of these plants, while a little tough, are still edible and can be used in the same way you'd use cabbage leaves in cooking.

If broccoli and cauliflower heads are left on the plant too long, they will begin to flower. The flowers are still edible. If cabbage heads are left for a long time, they can split and become unusable.

Herbs: Many herbs, like sage, thyme, tarragon, and oregano, are perennial and will not only grow throughout the summer but will return again next year. Just trim off the top ⅓ or so of the stems when you need some herbs. Even if you don't need them, it's still a good idea to trim periodically because if herbs are allowed to flower, they'll stop growing until the flowers are removed. *Annual* herbs like parsley and cilantro can be harvested the same way.

DEFINITION

An **annual** is a plant that completes its lifecycle within a one-year period and must be replanted each year. However, if an annual is allowed to flower and drop its seeds, new plants will grow the next year.

Harvesting Root Vegetables

Harvesting root crops is one of gardening's great pleasures because there's always a little element of the unknown. Even if you think you know what to expect, there's usually something about the vegetable's size or shape to surprise you. Plus, some root vegetables offer a two-for-one because the greens are also edible.

Most root crops come out of the soil easily unless it has become incredibly compacted, but sometimes carrots get stuck and need to be dug out with a trowel. If you do attempt to harvest a vegetable and end up with nothing but a handful of greens, don't worry. If you don't want to go digging with a trowel, just wait—the root will develop a new set of greens, and you can use those to pull it out of the ground.

Beets: Beets are often harvested when they're about the size of a tennis ball, although some varieties are meant to grow larger. Many gardeners prefer their beets on the smaller side, when they're more tender and sweet. The beet roots begin to push up out of the soil as they grow, and that gives you an indication of the size of the beet. Beet greens are delicious and can be prepared in the same manner as chard

Planting beets always guarantees you some colorful microgreens for your salads because a beet "seed" is actually a small fruit consisting of a cluster of seeds, so it will send up multiple seedlings. These seedlings should be thinned, leaving one behind to grow into a beet root.

SMALL STEPS

Beets, turnips, and radishes can all still be eaten after they've bolted and flowered. However, the flower pulls energy from the plant root in order to grow, so the root will be smaller after the vegetable has bolted

Turnips: The guidelines for harvesting turnips are very similar to beets. The roots grow up above the soil, and somewhere between 2 or 3 inches in diameter is a good size for harvesting. Turnips that are allowed to grow too large can become woody. Turnip tops can be prepared as a spicy cooking green. Very young turnips can be harvested when they have first sprouted and used as spicy salad microgreens.

Radishes: The ideal size of the radish depends on the variety, but look for the top of the root above the soil line for clues on the size. They grow quickly, so don't neglect them for too long! Just like turnips, radish greens can be used for cooking, and little radish sprouts are delicious.

Carrots: Carrots can be a little tricky because their root tops don't always grow up above the soil to signal their size. It's possible, through experience, to judge the size of a carrot by the way the carrot greens are growing. If necessary, you can scoop your finger into the soil to uncover the top of the carrot and discern its size. It's usually a good idea to wait until the top of the carrot root is at least the diameter of a dime before pulling. Smaller carrots are sweeter than large carrots.

Recipes are available for carrot tops, but some people get a rash when their skin touches wet greens. Backyard goats *love* to eat carrot greens, though!

Potatoes: The best time to harvest potatoes is after the tops of the plants have turned brown and died. Begin at the edge of the bed, and gently dig inward. It's very easy to accidentally cut potatoes when shoveling them out of the ground (which is why many people prefer to grow them in containers—see Chapter 4). If you'd like, you can "steal" a few new potatoes from your bed earlier in the season, when the plants begin to flower.

ROAD BLOCK

Potatoes that have been exposed to sunlight while growing will develop green-ish skins. Green potatoes are toxic and should not be eaten. To prevent green potatoes, mound additional soil around the plants every couple of weeks while they're growing.

Bulb onions: Onions will let you know when they're ready for harvest because their tops turn yellow and topple over. Carefully pull or dig out the onions. In areas that experience lots of temperature fluctuations, onions may bolt and begin creating a flower. The flower should be removed unless you'd like to save onion seeds.

Garlic: Unlike onions, it's normal for some varieties of garlic to send up a flower stem. When this happens, just cut the stem about ½ inch above the plant leaves to redirect energy downward to the bulb. Harvest garlic when 50 percent of the leaves have turned brown.

Harvesting Summer Crops

While spring greens and root vegetables are delicious, it's often the summer crops that gardeners await with the most anticipation. Everything covered in this category is, botanically speaking, a fruit, so we tend to ask if it is "ripe" when determining if it's ready for harvest. Summer crops generally continue producing until they are put out of com-mission by cold weather. Removing the ripe fruit helps the plant reallocate energy toward growing new fruit, so regular harvesting is good for overall productivity.

Tomatoes: The question of when to pull tomatoes off the vine doesn't always have a clear answer and involves a bit of personal preference. Some gardeners let the tomatoes ripen fully on the vine and pick them right before eating. Other gardeners choose to pick toma-toes when they're still partially green and allow them to finish ripening on the windowsill or kitchen counter. As long as the tomatoes have started the ripening process, even if it's just a blush of red or yellow on the skin, they will fully ripen on their own if left at room temperature.

The advantages of picking tomatoes early include preventing fruit damage from environ-mental stressors and removing opportunities for pests and critters to eat the fruit. For tricks on ripening green tomatoes after the frost hits, see the "Extending the Harvest" section later in this chapter.

ROAD BLOCK

Uncut tomatoes, whether they're fully ripe or not, should not be stored in the refrigerator. Cold temperatures negatively alter both the flavor and the texture of tomatoes. Once you've cut a tomato, it's generally best to store the remainder in the refrigerator unless you're going to use it quickly.

Peppers: The best way to know when a pepper is ready to be harvested is to familiarize yourself with its intended size and color. When it matches the description on the seed packet, it's ready to go! Any kind of green pepper will usually be ready first; red or yellow peppers start green but take time to ripen. If you accidentally harvest a red or yellow pepper early, it isn't a problem. Peppers, like tomatoes, continue to ripen after they're picked if left out at room temperature. Or you can just eat it when it's green, like a green bell pepper. Some peppers usually harvested when green, like Anaheims, will turn red if left on the plant. Hot peppers tend to get hotter the longer they're allowed to grow.

Eggplant: Gardeners generally agree that eggplants should be harvested when they're still on the small side because they have a better flavor and are more tender when they're young. A reflective sheen on the skin of an eggplant is another indication that it's ready to be picked. Try to take eggplants while they're still firm; fruits that stay on the plant too long will become mushy. Be a little careful when harvesting eggplants; the fruits often have sharp thorns at the top near the stem.

Cucumbers: Cucumbers can be picked at pretty much any stage in their growth. You may find that younger cucumbers have slightly concave sides, and you might prefer to wait until they "fill out" before you harvest them. Don't let cucumbers stay on the vine too long, though—if they turn yellow, and they're not a yellow variety, they become bitter.

Summer squash: These can also be picked either when they're small and tender or later when they reach their intended size. Once again, a little smaller is better in terms of taste. If summer squash stay on the plant too long, they develop the hard skins and seed pockets of a winter squash. Use a knife to cut the stem of the squash from the plant. Pulling or twisting the squash can often result in tearing, which causes the squash to rot more quickly.

Winter squash: Winter squash give a few different signs to show they're ready for harvesting. While they're growing, their skin is soft, but the skin of mature winter squash becomes hard. The squash gradually assumes the color it's meant to be; pumpkins turn from green to orange, spaghetti squash turn from light yellow to deep yellow, etc.

Lastly, check the stem of the squash. Fruit that's still ripening has a green, alive-looking stem; fruit that is ready has a stem that's turning brown and woody. You should be able to easily break your mature winter squash from the vine. If the stem is still pliable and wants to hang on, the squash may not be ready yet.

Melons: Determining whether melons are ripe can be challenging, and there are a number of different ways to do it. Ripe melons should separate easily from their stem when gentle pressure is applied. The spot where the melon was lying on the ground should be yellow or cream-colored, not white or pale green, when the melon is ripe. Some gardeners like to give their melons a little tap and listen for a dull thumping sound to determine ripeness. Also, when the curly little vine that grows opposite the fruit turns brown, the melon is ripe.

Extending the Harvest

The joys of gardening, and eating homegrown vegetables, don't have to end with the arrival of cool weather. True, the first frost usually puts an end to your wonderful summer crops, and that puts most gardens—and gardeners—into hibernation. However, there are ways to prolong the life of your summer crops, plus lots of great vegetables that don't mind growing in the cold.

You can provide some protection from an upcoming frost by covering your garden beds at night with bedsheets or floating row covers. This prevents the damaging effects of the first light frosts. When temperatures eventually drop to the point that the plants freeze and die, you'll need more protection.

If you're a container gardener, dealing with frosts is one of the ways in which you have in-ground gardeners beat. Your summer crops can continue to grow well for a long time if you bring them in at night and put them outside again during the day. Eventually, though, daytime freezing temperature and precipitation will damage container plants if they aren't given additional protection.

When the Frost Comes

As mentioned in Chapter 5, you can find out your projected first frost date by inquiring with your local agriculture extension office. It's a good idea to learn this date ahead of time and make a habit of checking the temperature forecasts when the frost date is getting close. In other words, don't let the first frost catch you off guard. Frost will not only kill most of your summer plants—tomatoes, peppers, eggplant, summer squash, basil, cucumbers, melons—it will also ruin the fruits attached to the plants.

SMALL STEPS

If you check your local weather forecast, you'll probably find a frost warning given in addition to the high and low temperatures. A frost can occur anywhere between 36°F and 32°F, but many gardeners consider a projected low of 32°F to be their indication of an upcoming frost.

If you know the frost is coming, it's time to do a sweep of your garden. Pick all the summer squash, eggplant, and cucumbers, even if they're small. Peppers and any tomatoes that have started to ripen can be brought indoors to finish ripening on your counter. Unfortunately, melons do not ripen off the vine, so any that aren't ready by the first frost may become compost.

But what about the green tomatoes? Unless they have a touch of red or yellow on their skin, the ripening-on-the-counter trick won't work. However, there are two ways to ripen even the greenest tomatoes. The first is to pull out the entire plant, including its roots, bring it inside, and hang it upside down. The tomatoes will ripen as long as they stay attached to the plant. Some people hang their plants from the rafters; others have been known to use their shower curtain rod.

If this seems a little space-prohibitive to you, you can place the green tomatoes inside a paper bag along with a ripe apple or banana. The fruit releases ethylene gas, which helps the green tomato ripen.

Of course, some gardeners see a bounty of green tomatoes as a blessing, not a curse. Fried green tomatoes, anyone?

It's best to harvest all potatoes before the frost. If you can't get to them in time, they might still be okay. A light frost may not damage them, but multiple frosts or a deep freeze will turn the potatoes to mush.

Winter squash work pretty much the same way. If you can, bring the squash inside before the frost. If not, a light frost will kill the vine, but the squash can still be harvested. However, if you leave the squash out in the garden for too long, a freeze will ruin them. Partially ripe squash will continue to ripen after they're picked. To learn more about storing potatoes and squash through the winter, see "Root 'Cellaring'" in Chapter 16.

Crops That Grow Well in Cool Weather

Not everything in your garden has to be rushed inside when the weather turns cold. Most of the same vegetables that can be planted before the last frost in the spring will continue to be harvestable well into the fall. (See the garden planning chart in Appendix C.)

Kale tastes slightly sweeter after it's been exposed to frost. Root vegetables like beets, turnips, and carrots can be harvested through late fall (and sometimes into winter) if they are mulched heavily before the first frost. Salad greens like lettuce and spinach tolerate a certain amount of cold weather, but once they experience a hard freeze, they will die. Basil will get knocked to the ground by a frost, but perennial herbs like thyme can continue to be harvested even in the snow!

Just because these vegetables will live in your garden during cool weather doesn't mean they can be planted when it's cold. As I discussed in Chapter 7, most vegetable seeds need the soil temperature to be at around 77°F in order to germinate. You can use a cold frame to raise your soil's temperature and successfully grow new crops, but if you simply try to plant lettuce seeds in your garden, it's unlikely they'll germinate in cool weather.

Cold Frames

Cold frames make it possible to continue growing right through the winter. A conventional cold frame is a simple structure that's placed on top of a garden bed. It consists of four low walls that shelter the plants and help trap heat, plus a clear lid that lets in sunlight. The cold frame walls can be constructed out of any of the materials used to make raised beds. Because the walls of the cold frame won't actually be touching the plants or holding the growing soil, it's more acceptable to use pressure-treated wood if necessary. The most popular choice for a cold frame lid is an old window, but Plexiglas or plastic sheeting attached to a rectangular wooden frame can also function as a lid.

This cold frame is made from 100 percent repurposed materials.
(Courtesy of doorgarden.com)

You'll find that one of the biggest challenges in using a cold frame isn't keeping the plants warm enough; it's preventing them from getting *too* warm. Be sure your cold frame lid can be partially opened and fully removed. If the temperature outside climbs into the 40s, open the cold frame lid 6 inches. If the temperature reaches 50°F or higher, remove the lid completely. Re-cover the cold frame in the late afternoon to trap some heat and help the plants get through the night. To help the cold frame retain heat overnight, cover the lid with blankets, newspapers, or anything else that could act as insulation.

An alternative to a conventional cold frame is a tunnel-like structure built out of PVC pipe and plastic sheeting. This may be a less-expensive option if you don't have cold frame materials readily available. The PVC cold frame is basically a long, low tent settled on top of a garden bed. Once again, be sure there's adequate ventilation for warm days.

Some gardeners choose to construct a cold frame on top of an existing raised bed by bending heavy gauge wire into arcs, securing the arcs to the walls of the raised bed, and covering the bed with plastic sheeting.

Sharing or Selling Your Harvest

At some point, your garden may produce more food than you can consume or preserve. (See Chapter 16 for information about food preservation.) If so, you can consider donating your excess food. Many nonprofit organizations, including homeless shelters, safe houses, and food pantries, would love to receive your fresh vegetables. Often these groups keep a stock of nonperishable items on hand, but fresh produce is hard to come by and greatly appreciated.

You can also organize a produce swap among the gardeners in your neighborhood. With the limited space available in cities, it can be difficult for one person to grow a large variety of items. Arranging trades after harvesting—and even coordinating planting in the spring—is a way to enjoy a large variety with a limited amount of garden space.

Selling some of the extra produce is also an option, and it can be a great way to offset the cost of seeds and gardening supplies. However, both your state health department and your city's zoning codes will affect how, and whether, you're able to sell your vegetables. For example, many states say only produce grown by an "approved source" can be sold to restaurants or at farmers' markets, but what it means to be an approved source isn't always clear.

Some health departments allow the sale of uncut, unprocessed produce as long as the grower is using city water and following the package instructions when using fertilizers and pesticides. Other states require an inspection, a fee, and a certificate before growers are allowed to sell to the public or to restaurants.

In terms of zoning, your city might have restrictions against growing anything on your property that's for "commercial use" (that you intend to sell, even if it's only a little). Even if your city doesn't care whether your tomatoes end up on your plate or a customer's, they likely have restrictions on selling anything within a residential area. So setting up a little farmstand in front of your house wouldn't be acceptable. You can transport your produce to a business district to sell it or set up a table (by yourself or in partnership with other gardeners) at a local farmers' market.

It's always a good idea to check your local health department and zoning rules before making plans to sell your extra produce. You can also talk to the leaders of the sustainability groups in your city because they're likely up-to-date on your area's rules and restrictions.

SMALL STEPS

Some urban sustainable living groups are setting up minimarkets as a way for neighborhood gardeners to sell their extra produce. Sometimes referred to as a handmade/homemade market, they also offer crafts, jewelry, and other goodies. These minimarkets are a great way for gardeners to earn seed money without requiring the expense and setup of participating in a full-blown farmers' market.

The Least You Need to Know

- Most plants continue to produce food for an extended period of time if harvested properly.
- An upcoming frost will damage the heat-loving plants in your garden, but some of their fruits can be ripened indoors.
- You can continue gardening in cool weather if you've planted cold-tolerant vegetables or if you use a cold frame.
- You have the option of donating, trading, and/or selling the extra vegetables from your garden.

Raising Animals for Food

Part

3

Haven't you heard that chickens are the new dog? That may not be quite true, but food-producing animals are becoming more and more common in cities. Animal protein (in one form or another) is an important part of most people's diets. Purchasing eggs, milk, cheese, meat, and fish not only can be expensive, it can also be difficult to discern whether your choices are sustainable, healthy, and humane. Therefore, it makes sense that urban homesteaders would be interested in bringing the production of these items closer to home.

While your vision of what it means to raise animals for food might look like Old MacDonald's barnyard, in fact, it's possible to keep chickens, dwarf goats, rabbits, or fish in an urban setting. In addition, many homesteaders are discovering the joy of keeping a beehive or two—not just for the wonderful honey, but for the help the bees give in pollinating the garden.

In Part 3, I let you know how you can raise any (or all) of these animals successfully, right where you live!

Livestock in the City

Chapter

10

In This Chapter

- Animals best suited for the city
- Potential zoning issues you may encounter
- Tips for a successful backyard barnyard
- A basic animal first-aid kit
- Winning over your neighbors
- Who to go to for help

For those who are interested in bringing a little bit of the country life into the city, growing vegetables doesn't seem like such a stretch. But chickens? Bees?? Goats?!? Even though we willingly live alongside all types of urban pets, from 150-pound dogs to a 1-pound guinea pig, the notion of livestock in the city can feel like a stretch. However, for anyone who is sincerely interested in producing some of the food they eat, animals are a natural part of the equation. Having a garden is wonderful, but it's difficult to grow a significant amount of protein—especially in a small space.

Raising small animals to produce healthy, sustainable, and affordable eggs, dairy, meat, fish, and honey can absolutely be a part of urban homesteading. In this chapter, I talk about why livestock doesn't only belong in the country. I cover various aspects of creating a successful living space for your animals and tips for talking to your neighbors about your newest project. I go over how you can get help with your animals and arrange care for them when you're away.

City Slickers: How Urban Animals Are Different

It's entirely possible to raise food-producing animals safely, cleanly, and humanely in an urban setting. In fact, it's happening right now in cities all across the country (and around the world). The reason this can work is because there are a few significant differences between livestock raised in the city and livestock raised in the country.

SMALL STEPS

Want to keep food-producing animals in the city but don't have a yard? Each animal chapter in Part 3 contains a section titled "Raising (*type of animal*) Without Land" you won't want to miss.

The first difference is in the type of animals that urban homesteaders keep in their backyards. You won't find the larger livestock common in the country (no big cows and certainly no bulls!). Instead, city yards are home to smaller animals like chickens, goats, and rabbits. Even the goats are usually city size—most urban homesteaders choose to raise dwarf breeds. Dwarf goats weigh in at around 55 pounds, which is about the size of an average dog.

City animals tend to be those that can keep quiet and not act disruptively. What that usually means—at least in the case of chickens and goats—is that no boys are allowed. Roosters are not typically a part of backyard barnyards. They really do make a racket at sunrise—and they don't take weekends off!—plus they're not necessary for egg production. Female goats are gentle and clean, but reproductive males smell terrible and are sometimes aggressive. So while there are all kinds of livestock out on rural farms, only a few have what it takes to live in the city.

Another major difference between rural and city animals has to do with scale. Country farmers often raise hundreds, if not thousands, of livestock animals. In contrast, urban homesteaders only require a few animals to meet their needs. It's still important to carefully tend the animals, but there's no denying that there's a world of difference between maintaining a flock of 6 chickens versus trying to manage a henhouse with 2,000 birds.

None of these barnyard animals—chickens, goats, or rabbits—is inherently unclean. When managed properly, they're no more impactful on the neighborhood environment than a dog. In fact, livestock manure has the unique quality of being compostable and safe to use in gardens, while the feces from dogs and cats contains parasites that are dangerous for humans.

Urban homesteaders can also raise bees and fish, which follow the same rules of scale as the other animals. The typical homesteader keeps one or two beehives (as opposed to a couple hundred or thousand like commercial beekeepers) and just a small tank of fish.

Zoning Issues

If you live in a city, the zoning code will dictate pretty much everything about how, and if, you're allowed to keep food-producing animals on your property. If you're thinking of owning animals, check your zoning rules before doing anything else. I covered the basics of navigating—and changing—your city's zoning code in Chapter 2. If paging through the code is intimidating, you can begin your inquiry by calling the animal control department. It's likely that they're used to fielding questions about food-producing animals, so they may be able to advise you about your city's rules. Plus, your city's Animal Code may contain additional regulations regarding the keeping of animals that you'll want to be aware of.

ROAD BLOCK

City officials can sometimes give conflicting information regarding livestock— especially if you talk to people in different departments. Be persistent, read the code yourself, and talk with sustainable living advocates in your city to get a clear picture of the rules.

If your city does allow animals, it will almost certainly limit the type and number you're allowed to own. In addition, you're likely to run into some of the following conditions:

Space requirements: You may need to provide a certain square footage of living space per animal.

Lot size: Even if you can provide adequate space for your animals, there may be a minimum lot size requirement, such as 10,000 square feet, in order to have chickens. Rules like this are often used to effectively prohibit food-producing animals everywhere except the outskirts of the city.

Setbacks: Sometimes setbacks are easy to abide by, like stating that a chicken coop must be at least 5 feet from the property line. Other times, they're more difficult, such as requiring that all animals be kept at least 25 feet from any structure, including your own house or garage. Excessive setback requirements are another way to prevent anyone who lives on a standard city lot from owning livestock.

Location on lot: There are often rules meant to ensure that livestock isn't kept in the front yard. For example, the zoning code may state that the animals must be kept in the rear third of the lot.

Fencing: Sometimes fencing requirements are specified. This can be applied not just for chickens and goats but bees as well. Requiring that hives be kept behind a 6-foot fence is meant to ensure that the bees' flight path is overhead.

Predator proofing: Some codes specify the type of shelter you must provide and require that the animals be kept in a predator-proof enclosure at all times.

Disposal of manure: A variety of conditions can be placed on the disposal of manure. Some cities simply ask that the manure be dealt with in a way that doesn't create a nuisance for neighbors. Other cities ask that the manure be thrown away rather than composted, and sometimes the code specifically states that the manure cannot be used for fertilizer. (Can you believe that? What a waste!)

Breeding: Some city codes state that the livestock must be "nonreproductive." This is a problem if you want to keep dairy goats (who must be bred in order to produce milk) or if you want to raise rabbits for meat.

Permit: Your city may state that you need to get a permit in order to keep animals. The sky's the limit in terms of what might be required to secure and maintain your permit. The permit could be issued by the animal control or zoning departments (or both). A fee might be involved. Your permit might be a one-time thing, or you might need to renew it annually. You may have to submit to an inspection or provide proof of animal vaccinations or disease testing. A limit might be placed on the total number of livestock permits that can be issued in your city each year.

URBAN INFO

Just one example of how things can change: beekeeping was illegal in Vancouver until 2003. In 2010, beehives were put on the roof of City Hall!

Variance: Some cities ask that you apply for a zoning variance to keep animals. This is a long process that's required whenever someone wants to do something that deviates from the established zoning code. It involves public notice and a period for public comment. The variance may become a permanent part of the property record, which means that anyone who purchases your property in the future will be able to see that a variance was granted for the keeping of food-producing animals.

Neighbor permission: Some cities require that anyone interested in owning food-producing animals obtain written consent from neighbors. This is typically only applicable for adjacent neighbors, but sometimes there's a broader distance requirement, such as any neighbor living within 250 feet. Occasionally cities will only require that a portion of the neighbors give consent—for example, 80 percent of the neighbors living within a specified area.

"Barnyard" Basics

Each type of food-producing animal has different requirements for space and shelter. But whether they live in a coop, run, cage, shed, or fenced yard, if you keep your animals outside, you're going to need to pay special attention to how you care for them. Living in cities means living in close proximity with other people. It's important to take steps to ensure that your animals don't negatively impact your neighbors and that they fit in well with your larger community.

Choosing the Right Spot

Your first responsibility when selecting a living area for your animals is to ensure they're safe and comfortable. Chapters 11 through 15 address issues of space requirements, security, and protection from the elements for specific animals. Most urban yards (or roofs, or balconies) don't offer a lot of different options in terms of where you can put your animals. Your city's zoning regulations may further limit your choices. All that being said, here are some basic guidelines to keep in mind:

Know your neighbors. If your neighbor to the north is wary about chickens but your neighbor to south can't wait to share the fresh eggs, put the chickens on the south side of your yard. If your yard is bordered by an alley to the rear, that may be your best option.

Keep the housing structure as far from the borders of your property as possible. Even if your animals are allowed to roam in an open yard, keep their coop/shed/cage as close to the center of the yard as you can.

Try to keep the animals close to your house and preferably near an open window. That way, if they're in distress or creating a ruckus, you're more likely to notice.

Aim for privacy. Your animals are less at risk if it's not obvious to other people that you have them. If you don't have a solid fence that blocks your animals from the sight of passersby, consider building one or planting tall shrubs to make a screen.

If possible, keep your compost pile(s) near your animal area to facilitate easy recycling of the animal waste into fertilizer.

If you're raising bees, take care to ensure that the opening to the beehives is not pointing toward your neighbors or facing a commonly used walkway.

Creating a Combined Barnyard

The conventional way to raise animals is to keep every species separate, and if you had enough room, and enough materials to build multiple fences, that might be the way to go. However, most urban homesteaders consider themselves lucky if they have just one small yard space to call their own, so multiple barnyards aren't really an option.

Luckily, animals can share the space you have. Chickens and goats are the ones best suited for pairing. They should be given separate sleeping quarters—chickens must be secured at night, while goats don't always have to be—but they can run around during the day together.

This urban barnyard accommodates chickens and goats. The shed on the left serves as the goats' shelter and also holds the laying nests for the chickens. The chicken food is placed on top of the coop on the right, out of the goats' reach.
(Courtesy of BrianKraft.com)

The biggest issue is that the chicken food will need to be kept out of reach of the goats, who would like nothing more than to devour it. Luckily, chickens can fly and goats can't—although they can be quite the acrobats! A 4-foot-tall structure should be low enough for the chickens to fly up and get to their food but high enough that dwarf goats can't access it.

Some animal owners are able to secure their chickens' food in an enclosure that has an opening so small the goats can't get in. However, there are just as many instances of goats figuring out how to get into seemingly impossible spaces!

Another option would be to scatter the chicken's food (as long as it's not medicated) throughout the barnyard. They can hunt and peck for their lunch, but the goats won't be able to locate very much of it.

ROAD BLOCK

If the feed you're giving your chickens is medicated, it's *critical* that you not allow any other animal—goat, rabbit, dog, etc.—to get into the food. It could be fatal to them.

Goats are also known for climbing into chicken nests and breaking eggs, so try to keep them out of the laying area. There's always a possibility that a goat may accidentally injure a chicken during play, but as long as you have enough space for them, and as long as your goats don't have horns, things will probably be okay.

Putting chickens and rabbits together is a little trickier, but several homesteaders have made this combination work. In order for it to be successful, the chickens and rabbits need to have separate sleeping quarters, and it's important for the rabbits to always have access to their hutch so they can get away from the chickens when needed.

This kind of free-range situation is really only possible with a small number of rabbits. However, there are a number of potential problems with this scenario. Chickens can hurt rabbits and vice versa. It's very difficult to breed rabbits in shared quarters because the chickens are likely to eat the babies. It's difficult to keep rabbits clean if they're hopping through chicken droppings, and chickens and rabbits can pass on a number of diseases to each other. One method of space sharing that bypasses many of these concerns is to go vertical—suspending rabbit cages over the chicken area.

Letting rabbits and goats roam together isn't as problematic, although it would be quite challenging to keep the goats out of the rabbit food! The opposite problem could also occur because goats like to eat lots of fresh greens, but it's not a good idea to give too many of those to rabbits. Rabbits are easily startled and can become sick when stressed, and there's always the risk of a goat stepping on a rabbit.

In summary, you can safely and relatively easily keep your chicken and goats together, but it may be best to raise the rabbits separately. Beehives can be kept adjacent to your barnyard or in your garden. Fish, of course, prefer not to mix with their livestock brothers and sisters.

Keeping It Clean

In general, the more crowded your animals are, the more frequently you'll need to clean. The best way to catch the waste from chickens, goats, and rabbits is to provide them with some sort of bedding. Pine shavings and straw are both good choices, although straw is more economical for covering larger areas. The bedding makes it quick and easy to rake out the animal's living area, and the manure and bedding are easily compostable. If you're in a hurry, you can "spot rake" any soiled places or cover them with fresh bedding, and your animal area will look as good as new!

The moisture in animal waste is what causes things to smell. This will become quickly apparent when it rains and any odors that might be present become more pronounced. All-natural products like Sweet PDZ and Stall Dry are available, and you can use them to absorb moisture and the odor that comes with it. The products are not meant to be a substitute for good barnyard maintenance, but they can help provide an extra boost of freshness.

Keeping It Quiet

Food-producing animals are not an excessively noisy bunch. Rabbits, fish, and bees are all silent. Hen chickens and goats may make a little noise occasionally, but it's quieter than the average dog bark. Also, unlike dogs, chickens and goats don't make noise when they feel threatened—their response to a perceived threat is to become very quiet.

The only reason these animals would become disruptive is if their needs aren't being met. If they're made to go too long without food or water or, in the case of dairy goats, if they have to wait too long to be milked, they might start fussing. If you keep a fairly regular schedule and be sure your animals are well cared for, noise shouldn't be an issue.

ROAD BLOCK

Goats are very intelligent and will quickly figure you out. Be sure you don't get tricked into giving them what they want every time they make a little noise, or you'll soon have a problem on your hands. (See the "Food and Water" section in Chapter 12 for more information on spoiling goats.)

Animal First-Aid Kit

It's a good idea to keep a simple first-aid kit handy. You probably won't use it very often, but if problems do arise, you'll have the supplies you need. The same basic items can be used to care for chickens, goats, and rabbits.

Your first-aid kit should contain the following:

- Betadine or Blu-Kote for cleaning wounds
- Baby shampoo or castille soap for cleaning feathers or fur around a wound
- Baking soda, cornstarch, Wonder Dust, or Kwik Stop for small bleeding wounds
- Food-grade diatomaceous earth for larger bleeding wounds
- Petroleum jelly for irritated skin and protection from frostbite
- Needleless syringe for giving liquids to a sick animal
- Quik Chik or other electrolyte solution

SMALL STEPS

You can make a homemade electrolyte solution by combining 2 teaspoons salt, 1 teaspoon baking soda, and 2 tablespoons sugar in 1 quart water.

- Sterile eye wash
- Antibacterial ointment
- Digital thermometer
- Scalpel for lancing abscesses
- Medical gloves
- Activated charcoal gel (only if you have goats)

Some of these items are available in drugstores, but you'll need to purchase others online or from a veterinarian supply company.

Making Friends with the Neighbors

Urban homesteaders are often surprised at how excited their neighbors are about the idea of food-producing animals living next door. The opportunity to spend time with livestock can be a treat for kids and adults alike, and you may find yourself with an abundance of chicken-sitting offers.

However, even though food-producing animals are becoming more and more common in cities, there are still those who feel that chickens and goats "belong" in the country. Even if your neighbors aren't necessarily opposed to the idea of you keeping animals for food, they're likely to have some questions and concerns. The best approach to take is to be as open and communicative with your neighbors as possible. Be proactive! In most cases, your neighbors will respond more positively to your animals if they've had a chance to learn about them before they arrive.

Homesteaders tend to be a self-reliant sort, and because of that, it can be difficult to accept that what you choose to do on your property should concern your neighbors at all. (Translation: It's hard to understand why it's any of their business!) Unfortunately, if you live in the city, close neighbors are a fact of life, and some of them are bound to take an interest if a couple of goats show up next door. Try not to be defensive or resentful if you're approached by a neighbor who has questions about your animals. To borrow an old country phrase, remember that you always catch more flies with honey than with vinegar!

Be sure you have a plan for how you're going to care for your animals, including waste, noise, rodents, and predators. These are the things your neighbors are most likely to be concerned about. Remember that people who live in cities are usually totally unfamiliar with livestock, and their fears are often based on misconceptions.

They may not realize that your dwarf goat will be smaller than their dog or that your bees won't all congregate in their backyard. They have no idea of how food-producing animals could be kept in a city. It's your role to educate them. If you want a little help, go to eatwhereUlive.com and click on "Myths Versus Facts About Food-Producing Animals." Address the issue with your neighbors respectfully, and they'll probably end up being supportive. This is not to say that there aren't some truly irrational people in the world (there are!), but most folks have the capacity to be reasonable if their concerns are addressed.

The last tip for winning over your neighbors is probably the simplest: bribery works. Or to put it more diplomatically: you can share the abundance of your harvest with them. Nothing will go further toward convincing your neighbor that urban chickens are beneficial than supplying him or her with fresh backyard eggs. It's very likely that your animals will provide you with more milk, honey, or eggs than you truly need. Even if you can technically use it all, it's still a good idea to share. A little generosity goes a long way.

Fitting Animal Care into a Busy Life

There's no doubt about it—everyone is busy these days. Even with an active life, you can still raise food-producing animals. Most of them don't require large amounts of time, just consistency. The consistency is important, though, because without it, your animals won't be healthy. If you don't think you have time to care for a dog or cat, you probably shouldn't get livestock.

Each of the chapters in Part 3 includes an estimate of how much time you should expect to spend caring for each type of food-producing animal. The responsibilities of caring for these animals are real, but the rewards are many. Most homesteaders find that the joys of food, fertilizer, entertainment, and companionship that they can get from their animals are well worth the work.

If you want to raise food-producing animals but feel you'll need some help, consider finding a partner. Join with a friend, or even a network of neighbors, who wants to share in the work in exchange for a portion of the harvest. Rotating care is especially helpful when faced with the milking schedule for dairy goats. You can create a kind of collective with a shared milking schedule. This not only helps distribute the work, but gives you the flexibility to swap shifts when needed.

One of the main benefits to homesteading in the city is the wonderful element of community support. People who live in the country don't always have a neighbor living a stone's throw away, but you probably have hundreds of neighbors living within a block or two of you! Invest a little time training a few willing neighbors to take care of your animals, and you'll have a sitter (plus a couple of backups) whenever you go on vacation.

If you need to find additional options for animal care, try reaching out to the sustainable living community in your city. The number of professional chicken-sitting businesses in urban areas is growing, and these folks are more than happy to care for your flock while you're away.

SMALL STEPS

Do you have a little extra time and an entrepreneurial spirit? Start your own chicken- (or goat- or bee-) sitting business! You'll have very little overhead, and you'll be on the cusp of the urban livestock movement.

The Least You Need to Know

- Not all livestock is created equal; some work better in the city than others.
- Even if your city allows food-producing animals, you're likely to face a number of zoning restrictions or regulations.
- Although your neighbors may not be wild about living next to livestock, you can win them over if you respectfully address their concerns.
- If you need help raising or caring for your animals from time to time, it's available.

Chickens Coming Home to Roost

In This Chapter

- Selecting the right chickens for you
- Caring for baby chicks
- Housing and feeding chickens
- Predators and other challenges
- Collecting eggs and raising birds for meat

A few minutes spent watching chickens leisurely scratch and peck can be the perfect way to counteract a stressful day. Chickens are extraordinarily well suited for urban environments because they don't need a lot of room and don't make a lot of noise. Instead, they simply produce wonderfully healthy and tasty eggs—and fantastic garden fertilizer, too! If you'd like to raise chickens in the city, you've come to the right place.

In this chapter, I talk about how to decide which chickens are right for you. I go over the basics of caring for chicks and adult chickens and address some of the challenges you may encounter. I also discuss the egg-laying habits of backyard birds and how to proceed if you'd like to raise chickens for meat.

Choosing Chickens for the City

Urban homesteaders have a wealth of chicken varieties to choose from. Birds are available in a wide spectrum of sizes, colors, and temperaments. You can choose from birds with puffballs on their heads, feathers on their feet, or naked necks. Some birds are lean and productive, and others are fluffy and "lazy" (at least when it comes to laying eggs) but are great with kids. Every homesteader has different needs and different things he or she is hoping to receive from the flock. Creating your own custom-designed flock is one of the joys of urban homesteading.

URBAN INFO

While chickens are quite popular, they're not the only kind of poultry you can keep in your urban backyard. Ducks, geese, and even turkeys have also been successfully raised in cities.

Despite all the choices, there's one decision that's usually non-negotiable for urban dwellers. Roosters (male chickens) have qualities that make them undesirable for city backyards. The caricature of a rooster crowing with the sunrise is solidly based in fact, and your neighbors may **not appreciate** the early morning wake up call.

Layers or Meat Birds?

The lure of farm-fresh eggs is what draws most people into owning chickens. Chickens raised for eggs are called *layers*. Layers typically reside in a backyard barnyard for at least a couple of years and are sometimes considered pets.

Even though layers are often the first choice for new chicken owners, it's also possible to raise chickens for meat—even in the city. Breeds intended for meat are called meat birds. Meat birds grow very rapidly and are usually "harvested" (that is, slaughtered) at a young age: between 2 and 4 months old, depending on how you plan to cook the chicken. Raising meat birds is more of a short-term project compared to layers.

DEFINITION

A **layer** is a chicken that's been bred for its egg-laying qualities and is raised for egg production.

Chicken owners looking for the best of both worlds can select a dual-purpose breed. These birds can be used for either eggs or meat.

Recommended Chicken Breeds	
Layers	White Leghorn, Golden Comet, Red Star, Black Star, Rhode Island Red, Australorp, Dominique, Barred Rock
Meat Birds	Cornish Cross*
Dual-Purpose	Rhode Island Red, Wyandottes, Dominique, Orpingtons, New Hampshire Red, Plymouth Rock

Cornish Crosses are hybrid birds bred to gain weight very quickly. Because of this, they're often the preferred choice for meat birds. However, the ultrafast weight gain can cause a multitude of problems for the birds, and some chicken owners consider raising Cornish Crosses inhumane. The alternative is to raise a dual-purpose breed for meat that doesn't gain weight quite as quickly but doesn't have the problems of a Cornish Cross.

Other Considerations

Deciding on your preferred chicken breed (or breeds) is only the beginning of the process. There are a number of additional factors to consider when choosing your chickens:

Number: The number of chickens you raise should be determined, first and foremost, by the amount of space you can provide for them. See the following "Chick Care" and "Chicken Care" sections to learn about the space required to maintain healthy birds.

After that, you should decide how many eggs you'd like to receive from your flock. In general, you can expect two eggs per three hens each day. This number will be slightly higher when the chickens are younger, and it will decrease during winter months and as the chickens get older.

Remember, especially if you're raising the birds from baby chicks, that you might want to order an extra bird or two to allow for the possibility that they may not all survive to adulthood. There's also a chance you'll accidentally receive a male chick instead of a female, and you'll need to find a new home for the young rooster. Just be sure you can accommodate all the birds you order in case they all grow up to be healthy hens!

SMALL STEPS

Taking care of chickens isn't time-consuming. You need to check on baby chicks several times a day, but adult chickens are easier. They need about five minutes each morning and evening, plus a little extra time each week for cleaning.

Temperament: Chickens have individual dispositions, and it's possible for a hen whose breed has an unruly reputation to end up being a great backyard bird. However, when chickens will be raised in the close quarters of an urban yard, choosing breeds known to be docile is good for the well-being of the overall flock. If young children will be participating in the care of the chickens, easy-to-handle breeds will make the experience more pleasant. Brahma, Orpington, Delaware, Houdan, New Hampshire Red, and Plymouth Rock are breeds known to have a calm temperament.

Size: Chickens are available in two basic sizes. The first are what you think of as regular-size chickens. They're technically called large breeds but are sometimes referred to as "standard." These chickens generally weigh between 4 and 7 pounds. Chickens are also available in bantam breeds. Bantam chickens are small, weighing between 1 and 2 pounds. It might seem that smaller chickens would be better for city backyards, but that's not always the case. Bantam birds require just as much care as a large bird, but the eggs they produce are quite small. Also, bantam birds are more of a flight risk because of their lighter bodies.

Age: If you're raising meat birds, you'll likely start with baby chicks or very young birds. Meat birds generally don't live past four or five months, so there aren't many older birds to choose from!

If you're raising layers, you have the option of purchasing baby chicks, pullets, or adult hens. A pullet is any hen that's less than a year old, but the term is usually applied to a young hen who has just started laying. Baby chicks require a little bit of extra work. (See the later "Chick Care" section.) However, buying hens at that age has a number of advantages, including the wide selection when ordering baby chicks and the fact that the birds will likely be easier to handle as adults if they're raised from young chicks. Purchasing pullets allows you to bypass the work of chick raising, but your options for finding different breeds becomes limited, and you need to be sure the birds you're purchasing are not ill.

If you buy an adult hen, you run the risk of accidentally getting a bird that's older and no longer laying. Chickens typically only lay regularly for two or three years. Young birds that are still laying will have a moist, pink vent and a soft, flexible breastbone. (The vent is the opening on the chicken's backside where they emit eggs and droppings.)

Egg color: Each breed of chicken lays eggs of a particular color. They can be anything from white to brown to green to pink, depending on the breed. There won't be a difference in taste or nutritional content among the various eggs of your birds—although all the backyard eggs will be much tastier and more nutritious than what you can buy in the supermarket. However, it can be a lot of fun to collect and eat different-colored eggs, and it's also a bonus if you're hoping to sell your excess eggs.

URBAN INFO

A chicken's earlobe, the little flap on the side of its head, gives you a clue as to what color eggs the chicken lays. If a hen has white earlobes, she lays white eggs. If she has red earlobes, she lays brown eggs … unless she also has olive green legs. In that case, she lays green or blue eggs!

Where to Purchase Chickens

You have several options when it's time to purchase your chickens. A quick Internet search that includes your state plus the words "poultry," "chicken," and "association" will likely take you to a website where you can locate chicken breeders in your area. Depending on the time of year, you can get chicks, pullets, or even fertilized eggs from local breeders.

Breeders also sell their chickens on Craigslist, and some cities even have urban homesteading or gardening groups that advertise adoptable chickens on their online message boards. Local poultry swaps and stock shows are additional options for acquiring chickens.

It's important to avoid buying a sick chicken because one sick chicken can infect your whole flock. For that reason, it's a good idea to purchase birds from someone who has aligned themselves with a local poultry association. Signs of chicken illness include dull eyes, parasites under the wings or around the vent, and coughing or sneezing.

Although you might not have a feed store right down the block from you, there's probably one located just a little ways out of town, and they also sell baby chicks in the spring.

Another way to purchase baby chicks is through an online hatchery. Hatcheries offer a wide variety of chicken breeds and typically have a reputation for supplying healthy birds.

Right before chicks hatch they absorb the last of the yolk, and that sustains them for a couple days. This is what allows hatcheries to safely ship day-old chicks. Many hatcheries require a minimum order of 25 chicks because a large group is needed for warmth, but some are able to ship as few as 3 chicks at a time because they include a heating element in the package.

Some urban homesteaders coordinate orders in the spring, so it's possible to split a 25-chick order among 3 or 4 people. Even if you order all female chicks, be aware that the sexing process is only 90 percent accurate. However, if you order a "sex-link" variety, the males and females hatch out different colors, so sexing becomes easy. Red Stars and Black Stars are sex-link chickens.

Chick Care

Raising baby chicks can be an intense—but joyful—experience. The chicks don't actually require a lot of hands-on time, but most new chicken owners spend a lot of time just checking on the little ones.

Starting with chickens at the chick stage allows the birds to bond with you. That may seem like an irrelevant, touchy-feely sort of thing, but it can actually be quite helpful because it makes the bird easier to manage for the duration of its life. In addition, new chicken owners get to "warm up" with little, easy-to-handle chicks before needing to deal with a fully grown chicken.

Chick Housing

Baby chicks need to be raised in a sheltered, heated space. This can be a bathroom, a basement or attic, or a heated garage. You can put the chicks in a corner of your living space, but be aware that they create dust when they scratch at their bedding.

The box or container you use for holding your chicks is called a brooder. They're available commercially, but you can just as easily make your own. Brooders must be adequately large (at least 1 square foot per bird, but 2 square feet or more per bird is better), shelter the chicks from drafts, provide ventilation, and be secure. Pine shavings make great bedding material. You'll want to have your brooder completely set up and heated before your chicks arrive, so they don't have to wait any longer than necessary to get settled in their new home.

It's a good idea to cover the pine shavings with newspaper for the first day or two the chicks are in the brooder. Baby chicks can't always tell the difference between their food and the shavings, so a layer of newspaper over the bedding prevents them from filling up on shavings instead of food. Remove the newspaper after a couple of days because the slippery surface can cause abnormal leg development.

SMALL STEPS

A cardboard box that's used to ship a refrigerator can make a great brooder. Just lay the box on its side, cut off the top, and secure the opening with chicken wire.

It's important to provide the chicks with adequate heat. You will need an adjustable light with a 250-watt bulb. A red bulb is recommended because it prevents the chicks from picking at each other (more on this in the "Chick Challenges" section), but a white bulb will work if necessary.

Rather than worrying about achieving a precise temperature inside the brooder, start by positioning the light 18 inches above the bedding and watch the chicks' behavior. If they cuddle together tightly under the light, they're too cold and you should lower the light. If they spread out to the edges of the brooder, open their wings, and lie down, they're too hot and you should raise the light. As the chicks grow, they need less heat, so raise the light a little each week.

Food and Water for Chicks

Teaching chicks to drink is the most important task for a new chicken owner. Take each chick out of the shipping box and dip its beak briefly in the water dish *before* allowing it to run around the brooder. Watch the chicks carefully to be sure they all figure out how to drink. Add a little sugar to their water (3 tablespoons per quart) for the first two or three days to give the chicks extra energy. If you'd like, you can also add electrolyte powder like Quik Chik to the water.

Once you're satisfied that the chicks know how to drink, pick up each one and gently tap her beak on the food to prompt her to eat. Inexpensive chick feeders and waterers are available through online hatcheries, at feed stores, and possibly through Craigslist or Freecycle.

Laying chicks receive a starter feed for the first eight weeks of their life; meat bird chicks, for the first two weeks. After two weeks, meat bird chicks switch to a grower feed. (See the "Food and Water for Chickens" section later in this chapter.) Chick starter is available in medicated and nonmedicated forms. Medicated chick starter helps protect chicks from coccidiosis, a parasitic disease that's the most common cause of death in young birds. Chicks are more likely to suffer from coccidiosis when manure is allowed to accumulate in the food or water dishes, when the brooder is overcrowded, or when the bedding is wet. However, even if you take excellent care to keep the brooder clean, the chicks can still become sick.

Alternately, some hatcheries offer to vaccinate day-old chicks against coccidiosis before shipping, but others feel that the vaccination stresses the chicks and makes travel more difficult. Since laying chicks only consume medicated starter for the first 8 weeks of their lives and don't start laying until 19 or 20 weeks of age, the medication won't contaminate the eggs. However, droppings from medicated chicks should not be used for fertilizer, and care should be taken to keep other pets from eating medicated chick feed. If chicks are fed a commercially available chick starter, they don't need supplemental grit (which helps with digestion) because it's already included in the feed.

It's best not to feed produce to young chicks because it can cause digestive issues and loose droppings. However, chicks love to eat worms! If you're able to dig a few out of your garden and offer them to your chicks, they will be overjoyed.

Chick Challenges

If you're able to keep your brooder clean and provide your chicks with a constant supply of food and water, you're well on your way to having healthy chicks. Be sure to remove any wet bedding as quickly as possible because that's one of the conditions that can create illness in a flock.

Even healthy chicks will occasionally develop a condition known as pasting, in which the vent gets pasted over with dried droppings. Pasting can be fatal if it's not corrected. It's important to take a couple of minutes every day to check each chick's vent and be sure pasting hasn't occurred. If it has, take a damp paper towel and wipe off the droppings blocking the vent. Sometimes the droppings dry to the point that they have to be gently pulled off. It's not a fun process for the chicks, but it's worth it to keep them healthy!

Chickens really do create a pecking order, and it starts when they're chicks. A certain amount of roughhousing is expected, but occasionally chicks will pick at each other to the point of removing feathers or drawing blood. Picking is more likely to occur in a brooder that's too crowded or too hot. If you notice that picking is becoming a problem, reduce the heat in the brooder and give the chicks more space if possible.

At this point, it would be important to switch to a red lightbulb in the heat lamp because red will mask the appearance of blood. If chicks realize they've drawn blood, they're more likely to continue picking. It's very important to immediately remove any chicks that have been wounded and raise them separately until they have healed.

SMALL STEPS

Boredom can also cause chicks to pick at each other. Providing chicks with entertaining objects like coins, mirrors, balls, or empty toilet paper tubes can help prevent picking.

Chicken Care

Once your chicks get to be big enough, it's time to move them outside. The precise timing of the move should depend on the outside temperature, but it's usually around four weeks for meat birds and eight weeks for layers. The chicks need to be "feathered out," which means they've gotten enough of their adult feathers to help protect them from the elements.

If the temperature is dropping below 40°F at night when you first move the chicks outdoors, provide supplemental heat in the coop until they're at least 12 weeks old. (See the later "Chicken Challenges" section for information on supplemental heat.)

Chicken Housing and Predators

The first part of housing for chickens is the coop. A coop is the enclosure where the chickens sleep. In many cases, the coop is accessible to the chickens during the day, and it also contains their nesting boxes, water, and food. If the chickens cannot access their coop during the day, they need to have some sort of daytime shelter to house their nesting boxes and provide protection in case of extreme weather.

Coops do not need to be very large, especially if they're only used for sleeping. However, it's very important that the coop be completely predator proof. Chickens have limited night vision, so they're basically "sitting ducks" for any predator that's able to access them. Many predators will try to dig their way into a coop, so coops that are resting on dirt should have a hardware cloth barrier (or other protective material) buried at least a foot deep around the perimeter of the coop. Hardware cloth is a sturdy wire mesh material available at hardware stores. It is stronger than chicken wire. Coops need to have ventilation for the chickens, which is often accomplished with a secure, mesh-covered window.

Chicken coops can (and have) been built out of pretty much anything. A multitude of coop plans are available online. (See Appendix B for resources.) If you're looking for ways to build your coop cheaply, you can often find used and scrap materials on Craigslist and Freecycle.

URBAN INFO

A general guideline for nesting boxes is to have one box per two chickens. However, sometimes all the hens decide on a favorite box and will wait their turn to lay even when multiple empty boxes are available! Laying boxes can be built as part of the coop, but stacked milk crates also work well. The best way to teach your hens to lay in their nesting boxes is to put a few golf balls in there, to get them started!

Chickens sleep in their coops at night, but they need additional space to roam during the day. One way to provide this is with a chicken wire–covered enclosure called a run. A run is a way for chickens to enjoy the outdoors while still remaining contained. Sometimes the run is attached to the coop so the chickens can move freely between their inside and outside spaces during the day. Unless your run is also predator proof, however, you still need to secure your chickens in their coop every night and let them out each morning.

If you need to have flexibility in your schedule, invest in a predator-proof run. By securing hardware cloth to the floor of the run or burying it a foot deep around the perimeter, your chickens will be protected at all times during the day and night. If you have raccoons in your area, you may want to reinforce all six sides of the run with hardware cloth. As long as they have food and water, they can be left alone for a couple of days at a time without problems.

Providing access to an open yard is another way to give your chickens the daytime space they need. The yard must have a fence that's at least 4 feet high to keep the chickens in. Chickens can leap/fly up to 4 feet high, but a 4-foot fence is generally adequate to convince them to stay in their yard.

Whether your chickens spend their day in a run or a yard, they must have *at least* 4 square feet of space per bird. The more you can provide, the better for the chickens. Larger spaces are easier to keep clean because manure builds up quickly when chickens live in a small space.

Providing good bedding materials also helps keep the space—and the chickens—clean. Pine shavings and straw are both good choices, although straw is more affordable if you're providing bedding for a barnyard or large run. The chickens will scratch in the bedding and cover their droppings, which reduces odor. Also, the bedding material combined with the droppings creates a fabulous soil amendment for your garden!

ROAD BLOCK

Generally speaking, chickens and gardens don't mix. There are stories of using chickens as pest controllers in gardens, but more often the chickens destroy the plants along with the bugs! If your chickens are allowed to roam your yard, erect a fence to protect your garden.

The decision about whether to keep your chickens in an enclosed run or an open-air yard is a personal one and depends quite a bit on the predators that live in your neighborhood. Foxes, raccoons, coyotes, hawks, and dogs are all common urban predators. If you're concerned, you can purchase and apply coyote urine (available at sportsman's stores and online) to your fencing and coop as a way to deter predators. You can also use electric fencing to attempt to keep predators from getting into your yard.

Raising Chickens Without Land

Chickens are small, and they really don't need a lot of space. It is technically possible to raise a couple of hens indoors. If you get them a large enough cage, they will be comfortable, but cleaning the cage will likely be a daily chore. In addition, numerous websites offer chicken diapers for sale. (Seriously!)

Several other options are available to you if you'd like to raise (or help raise) chickens but don't have a yard. You can set up something similar to the gardening land-share agreement discussed in Chapter 6 but with chickens instead! Find someone who has a little land and is interested in owning chickens but who isn't able to do all the work him- or herself. You can agree to share the costs of creating a coop and ordering the birds and work out a division of labor for the chickens' daily care.

Or maybe someone's willing to front all the financial costs in exchange for you taking care of the chickens. The person would need to live pretty close to you in order for this to fit reasonably into your schedule, but if you have to walk a block or two to care for your chickens, you're probably not going any farther than a farmer does to get to his henhouse!

Another possibility is to find some unused land and establish a "community barnyard," similar to a community garden, with a co-op of interested homesteaders. Or you can find someone who already owns chickens and offer your services as his or her assistant. You can locate other chicken owners in your area through local sustainable living groups (there may even be one devoted specifically to chickens), by asking at the feed store, or by checking out your state's poultry association.

Food and Water for Chickens

Access to clean water is a critical element in the health of your flock. Chickens left without water, or whose water is allowed to become dirtied with droppings, are at a higher risk for illnesses. This is especially important during daylight hours. Some chicken owners don't keep water in the coop at night because the chickens generally don't move around when the sun is down. However, chickens need to have access to water in the morning, so if you're not confident of your ability to let the chickens out of their coop bright and early, it's better to keep water in the coop at night.

You'll need to have a plan to keep the water from freezing during the winter. You can purchase a stand for your waterer that heats up to prevent the water from freezing. There are also heated buckets or heated floaters you can put into a watering trough. A wide variety of waterers and watering accessories are available online or at your local feed store.

Layers consume chick starter until they're 8 weeks old, and then they can begin consuming a growing feed, sometimes called a flock raiser. They'll eat the growing feed until they're 19 weeks old, when they should be switched to a laying feed. Meat birds only eat chick starter for two weeks, and after that, they eat the growing feed until they're harvested.

You can purchase chicken feed, along with feeders for adult chickens, at a feed store. Also, sometimes local farmers mix their own feed and sell it through co-ops or on Craigslist. Commercially prepared feed—what you're likely to find at the feed store—has been very carefully formulated to give chickens all the nutrients they need to stay strong and healthy. Chickens that are fed commercial food should not need any kind of supplements, like grit or oyster shell.

Calcium is a critical element in laying feed. Without it, the chickens will begin to lay eggs with thin or missing shells. Limestone or oyster shell is often included in laying feed to provide the necessary calcium. Protein is important for the proper development of meat birds, so be sure their growing feed has at least 20 percent protein.

Although well-formulated feed provides all the necessary nutrients for your chickens, it's still nice to give them a treat once in a while. You can purchase chicken scratch, which is a combination of grains, to sprinkle on the ground for them periodically. It's nowhere near nutritionally complete and shouldn't make up a big portion of their diet, but small amounts are okay and give the chickens a chance to hunt and peck for their food.

Chickens are also incredible composters, and they will make short work of most of your kitchen scraps and garden waste. Part of what makes backyard eggs so delicious is their deeply colored yolks, which are caused by the beta-carotene they get from eating greens. So don't hesitate to feed your chickens your leftover salad and dandelion weeds!

The following table outlines some do's and don'ts when it comes to treating your chickens.

Chicken Treats

Yes	No
Fruits and vegetables	Raw green potato peels
Dairy products (especially yogurt)	Avocado skin or pit
Cooked meat and seafood	Dried or undercooked beans
Most garden scraps	Anything especially salty (like processed food)
	Moldy food
	Raw eggs or eggshells
	Onions or garlic
	Plants from the tobacco/nightshade family*
	Rhubarb leaves

Chickens can eat the fruits of nightshade plants (like tomatoes or eggplants) but not the plants themselves.

Chicken Challenges

Dealing with predators (see "Chicken Housing and Predators" earlier in this chapter) is usually the primary concern of urban chicken owners. However, additional challenges may arise:

Cold weather: Chickens are surprisingly resilient in the cold, but they often need a little help to stay healthy through the winter months. Once the temperature drops below 20°F, it's time to add supplemental heat to the coop. This can be done with a 60-watt lightbulb or a larger heat lamp—it depends on the size of your coop.

In addition, chickens with large combs (like Leghorns) need petroleum jelly applied to their combs during cold weather to prevent frostbite. If you live in an especially cold area, you should also cover the wattles of all your birds with the jelly for protection.

Molting: This typically occurs in the fall. Chickens may lose feathers for several weeks, and egg production will slow down or stop temporarily. Once the hens are finished molting, their eggs will be larger and egg production will increase.

Brooding: Brooding occurs when a hen decides to try to hatch chicks (even though her eggs aren't fertilized). This is a problem because the hen will stop laying when she's brooding. Plus, a broody hen rarely eats or drinks, which makes her more vulnerable to diseases. Allowing eggs to build up in a nest often triggers brooding, so it's a good idea to collect eggs daily. Heavier-bodied breeds are more prone to brooding. If you have a hen that's broody, do whatever you can, including blocking the entrance to the coop or covering the nesting box with a board, to prevent her from getting to the nest.

Escaping: Chickens who learn to fly over fences are at a greater risk of harm from predators—and traffic! Solutions include keeping chickens in a closed run, creating higher fences, or stringing a net over the top of the barnyard. You can "ground" your birds by carefully trimming the first 10 feathers of one wing with sharp shears. Be aware that wing clipping will leave your chickens more vulnerable if a predator gets into their yard.

Even with the best of care, sometimes chickens get injured or sick. Keep an animal first-aid kit on hand (see Chapter 10) to treat any injuries as they occur. It's also important to have a place, separate from your main flock, where you can keep a sick or injured bird until it's well. Chickens can catch any number of diseases. Even if you live in a highly populated urban area, you may be surprised to find that there's a veterinarian in your city who can care for chickens.

It's also immeasurably helpful to utilize online chicken forums because that allows you to get advice from veteran chicken owners. See Appendix B for suggested chicken forums.

Eggs and Meat

As I've mentioned, layers begin producing eggs at around 19 to 20 weeks of age. At first the eggs will be small and less frequent, but by 30 weeks, the hen should be producing well. A good hen will lay about 20 dozen eggs in her first year. She may lay six eggs per week during the summer, but egg production will drop significantly when the days grow shorter. Egg production declines as hens grow older until they only lay infrequently or stop laying altogether. Most chickens only produce eggs consistently for two or three years.

Check with your state Department of Agriculture to learn its rules regarding the sale of eggs. Many states allow you to sell eggs without any restrictions as long as you stay under a certain limit (such as 250 dozen per month). A small flock of six to eight hens can easily provide all the eggs for your family, and the sale of extra eggs to friends can pay for the chicken feed!

Chickens can live for 10 or more years if they're well cared for. Hens who are no longer laying can certainly be kept as pets, and they will continue to compost your table scraps into great fertilizer. However, urban homesteaders with limited space (and zoning rules that put a cap on the number of chickens you can own) are often faced with having to give up older chickens and replace them with a new flock. Some rural farms allow city dwellers to drop off older chickens, and the chickens can roam the grounds of the farm foraging for food. There may be someone on Craigslist who is happy to take your flock off your hands.

You also have the option of harvesting your layers. Two-year-old chickens are certainly not the best for eating, but they can be used for stocks or soups.

If you've raised meat birds, or you're planning on harvesting your layers, you'll need to find a place to do the butchering. Even the areas that allow chickens usually frown on butchering within the city. Use online searches, sustainability forums, and farmers' markets to find a small chicken producer in your area. You can also try to find a local farmer who butchers her own chickens. Offer to be her assistant the next time she harvests her birds, and she may let you process your chickens when she's finished with hers. If you don't want to harvest your chickens yourself, look for a shop that caters to hunters and is willing to process birds.

The Least You Need to Know

- Chickens can be raised for eggs or meat—or both—and certain breeds of each do best in the city.
- Raising a flock from baby chicks requires additional work but has numerous advantages.
- Providing chickens with clean water, proper food, and adequate space helps prevent illnesses and other problems.
- Predators are a fact of life in cities, but secure housing will protect your chickens from harm.

Getting Your (Dwarf) Goat

In This Chapter

* Selecting the right goats for you
* Goat care and challenges
* Breeding and milking dairy goats
* Raising goats for fiber or meat

Goats are about as close to an all-purpose animal as you can get. They provide delicious, healthy milk that can be used to make cheese, yogurt, ice cream, and soap. They can also be raised for fiber or harvested for meat. Goats thrive in a wide variety of climates and efficiently turn whatever you feed them into manure that can be used to enrich your garden. Goats are intelligent and curious and make wonderful companion animals. They're sometimes used as pack animals and enjoy going on hikes. City dwellers can even take their goat on a walk with a leash (but avoid any areas that may have dogs). In short, owning goats offers a wealth of benefits to urban homesteaders.

In this chapter, I talk about the different options you have when selecting a backyard goat. I go over the basics of providing food and shelter for your goats, as well as some of the challenges you may encounter. I discuss how to breed your goats for the purpose of producing milk, and I cover the steps for successful milking. I also touch on raising goats for fiber and meat.

Choosing Your Goats

Urban homesteading sometimes mimics country living—but things are shrunk down to a city-appropriate size. Urban homesteaders have smaller gardens and smaller barnyards. Luckily, city dwellers who want to experience the joys and challenges of raising goats can select a small breed that will live comfortably in an urban backyard. These wonderful goats are about the height and weight of a Golden Retriever. And while they may not play fetch, small goats provide a multitude of benefits for the people who raise them.

SMALL STEPS

Goats don't take a lot of time to care for. Basic care takes about 15 minutes a day, and you need to spend an additional 20 minutes each week to clean. If you decide to milk, plan on around 10 minutes per goat once you get the hang of milking, once or twice daily.

However, there are limitations on the types of miniature goats that are suitable for the city. *Does* (female goats) do well in backyards, but *bucks* (unneutered male goats) do not. Bucks have a strong, musky odor that makes them unpleasant to be around. The smell would certainly bother your neighbors—and most likely bother you, too! In addition, bucks have to be housed separately from does, which isn't practical on an urban homestead.

If you'd like, you can instead get a *wether*, which is a neutered male goat. Wethers are similar to does in size, and they don't smell bad the way bucks do. Wethers can come in handy when you're looking for an affordable companion for a doe.

It's important to have two goats because they're very social creatures and will literally become ill from stress and loneliness if they don't have a companion. Does require more of an investment, but wethers are usually cheap (and sometimes even free), so they can be a great way to round out your little herd.

Dairy, Fiber, or Meat?

Just as chickens raised for meat still lay eggs, and egg-layers can be used for soup, goats don't always fall into clear-cut categories. But within the world of small goats, there are certain breeds homesteaders prefer to use for one purpose over another.

SMALL STEPS

If you're planning on raising a goat for milk, be aware that she must have babies (and continue to be bred about once a year) in order to produce milk. Breeding, raising, and selling kids are all activities that can be successfully done in an urban setting, but it's good to know up front that dairy goats don't "automatically" produce milk for their owners!

Dairy: Nigerian Dwarfs are wonderful miniature dairy goats. They are known to produce an average of 1 quart of milk per day over a 300-day lactation period (the first 8 weeks of which are taken up by nursing their kids). The does weigh around 55 to 60 pounds (wethers are a bit larger) and are around 2 feet tall at the head when fully grown. They have a body configuration similar to full-size dairy goats, only smaller. The milk from Nigerian Dwarfs is richer (higher in butterfat) than other breeds. They don't have a specific breeding season but can be bred at any time, so if you own two does, you can stagger their breedings and have a year-round milk supply. Nigerian Dwarfs are most commonly raised for milk, but they're considered a dual-purpose breed and can also be used for meat.

As you can see, a fully grown Nigerian Dwarf doe is still pretty small, as far as livestock are concerned.
(Courtesy of BrianKraft.com)

URBAN INFO

Nigerian Dwarfs were the breed chosen for the Biosphere 2 experiment, in which scientists were sealed inside a self-sufficient dome for two years with the goal of creating a model for a space colony.

Fiber: Angora goats are renowned for producing beautiful mohair that can be used for clothing, carpet, yarn, and many other things. However, Angoras are full-size goats that don't fit well into city backyards. Urban homesteaders who want to raise a fiber goat can instead choose a Pygora, which is cross between an Angora and an African Pygmy. They're the same height as Nigerian Dwarfs and weigh around 65 to 70 pounds. These miniature goats can produce up to 2 pounds of fleece per shearing, and most are shorn once a year. Pygoras can also be milked (giving up to 1 quart per day) or used for meat.

Meat: African Pygmy goats are an option for homesteaders looking for a good meat goat. Pygmies are typically less than 2 feet tall and weigh 40 to 50 pounds. They have a stocky, compact build. African Pygmies can be milked, but their lactation period is only 130 days. They're able to breed year round, like a Nigerian Dwarf.

Registered Goats

A number of organizations register goats. The purpose of registering is to document the goat's lineage and prove that the goat is a purebred. If you're a homesteader simply looking for a healthy, well-producing animal, you may not care a bit whether your goat is registered or not. Registered goats often cost two or three times more than an unregistered goat, so buying an unregistered animal may be more practical. The only thing to keep in mind is if you'll be breeding your goat, you'll need to find homes for the kids. People interested in purchasing goats for 4-H projects, or who are otherwise interested in goat shows or goat genetics, usually only buy registered goats.

Purchasing Your Goats

The best way to purchase your goats, especially if you're a new goat owner, is to buy directly from a local breeder. Many local breeders advertise on Craigslist, and you can also find them through your state's goat associations. Goat shows and goat auctions don't always give you the time you need to ask questions and examine the goat, and you don't get a chance to inspect the goat's living environment. Evaluating how the goats are kept, and seeing the health of the other goats in the herd, tells you a lot about the animal you're considering purchasing.

Healthy goats should have a lustrous coat, a medium build (neither too fat nor too thin), and a spry gait. Be on the lookout for signs of illness, including runny nose or eyes, evidence of chronic diarrhea, patchy coat, or difficulty walking. The goats' living area should be reasonably sanitary and fresh smelling, with clean water and hay for the animals. It's best to buy goats who have had their horns removed (more on this in the later "Raising and Selling the Kids" section). You can expect to pay between $75 and $150 for an unregistered doe, $200 to $250 for a registered doe, and $75 or less for a wether.

Ask the breeder to provide documentation of the vaccinations and worming the goat has received. Some goat diseases are region specific; therefore, the recommended goat vaccinations will vary by area. You can check with your local agriculture extension or 4-H program to find out the best care practices where you live. You should also ask the breeder for documentation that his or her herd is free of a disease called caprine arthritis encephalitis (CAE). If you'd like to be especially thorough, you can ask to take the goat to a livestock veterinarian for a full examination before purchasing.

Goats can be purchased as young as 8 weeks old. Purchasing your goats as *kids* has some of the same advantages as buying baby chicks instead of grown chickens. Your goats may be easier to handle if you start with them as kids, and provided the kids are healthy when you purchase them, you'll be able to provide the care they need to stay that way throughout their life.

If you're purchasing dairy goats, it's important to inspect the mother as well as the kid. Ask the breeder how much milk she usually gets from the mother each day and ask to milk the mother (or ask the breeder to milk her while you watch).

A disadvantage to purchasing dairy goats as kids is that you cannot breed them right away. Although goats can be bred as early as 7 months old, it's advisable to wait until the goat is at least a year old. So while your new baby goats will be lots of fun to play with (and they'll provide you with great manure right away), it will be a while before they'll start providing you with milk.

You also have the option of buying an adult doe or even one who's already pregnant or lactating. Pregnant and lactating does cost more, but they do provide milk sooner. Inspect the doe's udder and teats. The udder should be well attached (not too saggy or pendulous) with one teat on each side. If the doe has been bred previously, ask the breeder about the *kidding* and the doe's milk production history.

 DEFINITION

A **kid** is a goat who's less than 1 year old. **Kidding** is the act of a goat giving birth.

Goat Care

Caring for small livestock in your backyard may seem like an intimidating proposition. However, once you begin to do it, you'll see it doesn't take a significant amount of time or effort, just a little bit of knowledge. Even though you don't *have* to spend lots of time with your goats, you may find that you *want* to. Goats are wonderfully curious, playful, and affectionate, and you'll learn that they enjoy your company at least as much as you do theirs. Goats rely on each other primarily for companionship, but regular human affection helps keep them happy and healthy.

Housing

Dwarf goats are not only easier to manage compared to a full-size goat, but they also require significantly less space. Goat housing must consist of a shelter and a fenced yard. The shelter should provide at least 15 square feet of space per animal and must be accessible to the goats all the time, although you may find that they hardly ever use it. It should be leakproof and contain at least a few inches of pine shavings or straw for bedding.

It's critical that the shelter be well ventilated because the ammonia fumes from soiled bedding will give goats respiratory problems. In fact, many goat owners opt to leave the shelter doors open at all times to ensure proper ventilation. Sheds make great backyard goat shelters.

Your goats will likely spend most of their time—except when it's raining or snowing—in their yard. The yard should provide at least 130 square feet of space per animal, although more is always better.

A fence that's 4 feet high should be adequate to keep in miniature goats. However, the fence needs to be strong and secure because the goats will most certainly rub up against it, headbutt it, and investigate it thoroughly for escape opportunities. Chicken wire is not sufficient for fencing in goats. Stock wire, sometimes called stock fencing, or chain link work well. Board fences are not preferable because goats can knock out a board and then get caught in the opening. Be sure your gate is especially strong, too, because the goats will likely focus their escape attempts there.

Goats do well on grass or bare dirt, but you may find that providing a layer of straw makes the yard easier to clean and maintain. Goats love climbing, so providing them with stacked straw bales or (solid) children's play equipment will be appreciated—just be sure you don't place it close to the fence!

Raising Goats Without Land

Technically, it's possible to house-train a goat. For one reason or another (often because a baby goat was ill or too small to survive on its own), there are many examples of people bringing goats indoors to live for a time. However, if you don't have experience with goats and at least the possibility of putting them outside when they get rowdy, the idea of keeping goats inside your home is pretty risky.

It's best to plan on raising your goats outdoors. Even if you don't own any land, plenty of other people in your city do. Many of the ideas I discussed in Chapter 11 for raising chickens without land would also be possible with goats. Find someone who has a yard and offer to share in the costs and care of the goats with them. Goats are great candidates for a "community barnyard" arrangement because they're sturdier than chickens and not as susceptible to predators. As long as goats have clean food, water, and are milked regularly, they're fine!

Because dairy goats require milking once or twice a day, it likely would not be difficult at all to convince a current goat owner to let you help with the work in exchange for a share of the milk. Some urban homesteaders acquire goats in their yard specifically with the intention of setting up a kind of milking co-op so the responsibilities of milking are shared by many people.

Food and Water

A goat's diet should consist primarily of hay. Hay is typically available in grass, alfalfa, or a blend of the two. Alfalfa is richer and provides more nutrients for pregnant and lactating goats, but it can cause other goats to gain weight.

URBAN INFO

Understanding the difference between hay and straw is something that's new to most urban homesteaders. Hay is grass or other plants (like alfalfa) that are dried and used as animal food. Straw is the dried stalks of cereal plants (like oats and wheat) after the edible parts have been removed. Straw is typically used for bedding.

Before purchasing your goats, talk with the breeder about what the animals are accustomed to eating. It's not a good idea to make sudden changes to your goats' diet. Feed them what they're used to when you first bring them home and gradually make any changes you feel necessary.

Goats are ruminants, and consistent access to roughage is important for keeping their rumens healthy, so try to be sure they always have hay. Goats will generally only eat as much hay as they need, so you don't have to worry about overfeeding.

You can get hay at a feed store, or you can find a local farmer who is offering his hay for sale. (They often advertise on Craigslist.) Good hay is green, fresh-smelling, and flaky. It should not be brown or moldy. Your feed store will probably sell feeders meant to hold hay, but for a couple goats, it's fine to just buy a bucket or two. You can also forgo any kind of container and just set the fresh hay in the corner of the goat pen, but some goats are too finicky to eat hay once it's touched the ground.

Grain-based goat feed mixtures are available to supplement your goats' diet. They provide certain nutrients that are helpful, especially for pregnant and lactating goats. Grains should be given to goats only in small amounts because too much will cause stomach problems and weight gain. Goats who are pregnant or lactating can receive a small amount of grain (around $\frac{1}{2}$ cup) twice a day; for other goats, it should be given sparingly as a treat.

Goats also benefit from a couple supplements that can be given "free choice," meaning they're left out for goats to eat whenever they like. Baking soda helps neutralize excessive acids in the stomach, and mineral salts provide vitamins and minerals to promote overall health. Feed stores carry goat feed and mineral salts, and you can also buy goat feed mixtures from local farmers.

Speaking of treats, goats will happily compost many kinds of table scraps and garden waste for you. Although goats have a reputation for eating anything (tin cans, anyone?), they can actually be quite picky and have delicate stomachs. Goats are herbivores, so plant-based foods are most appropriate for them. Also, any food that's grain based (like bread) should be given in limited quantities. Too much will cause the same problem as overfeeding grain mixture.

ROAD BLOCK

Goats are very easily spoiled, and spoiled goats are difficult to raise—especially in the city. Don't soothe your goat by providing a treat when it bleats, or the goat will very quickly learn to bleat … and bleat … and bleat … whenever it wants a treat.

The following table outlines some do's and don'ts when it comes to treating your goats.

Goat Treats

Yes	No
Vegetables and most fruits	Rotten food
Grain/bread products (in moderation)	Processed food
Most weeds	Stone fruits (peaches, cherries, plums, etc.)
Most garden plants	Meat or seafood
	Plants from the tobacco/nightshade family*
	Rhubarb leaves
	Raw green potato peels
	White clover
	Bindweed plants (like morning glories)
	Onions and garlic (milking does only)
	Pine needles (pregnant goats only—linked to miscarriage)
	Oak and wild cherry leaves

Goats can eat the fruits of nightshade plants (like tomatoes or eggplants) but not the plants themselves.

The main requirement for providing goats with water is to be sure it stays clean and doesn't run out. Goats generally don't go to the bathroom in their water the way chickens do, but they will sometimes drop straw and other barnyard debris into it. Goats will most certainly knock over their water if they're able to.

Feed stores sell flexible rubber tubs meant for feeding horses, and these work very well for goats because they're pretty much impossible to tip. Just as with chickens, make sure the goats' water doesn't freeze in the winter. See "Food and Water for Chickens" in Chapter 11 for winter water tips.

Maintenance and Challenges

As with all animals, an ounce of prevention is worth a pound of cure when it comes to keeping your goats healthy. Providing clean food and water, offering appropriate amounts of hay and grain, and keeping the yard clean are all simple ways to prevent problems. A few regular maintenance tasks are involved in goat care, as well as some additional things to be aware of.

Vaccinations and worming: Goats require annual vaccines. As I previously mentioned, the necessary vaccines vary from region to region, so check with your livestock veterinarian or your local agriculture extension for what you need. Goats should take a wormer twice a year. Chemical wormers are available, but many goat owners prefer the herbal wormer available from Hoegger (see Appendix B).

Hoof trimming: Goats also need regular hoof trimming. The frequency varies with the goat, but it's typically necessary about once every couple of months. You can ask your goat breeder to show you how to do it before you take your goats home, and you can buy the hoof trimming tool from a goat supply company (see Appendix B).

Winter care: Goats do well in cold weather, provided you keep the bedding in their shelter dry. A heat lamp is useful when the temperatures drop below 20°F or so. Do not close the shelter in an attempt to make it warmer—the lack of ventilation will cause more problems for the goats than the cold would.

Predators: Unless you live in an area with bears or mountain lions, your goats aren't in much danger from wild predators. Coyotes do exist in some cities and can potentially attack goats (especially goat kids), so if this is an issue in your area, you might want to use electric fencing for additional protection and coyote urine as a deterrent. The biggest threat to dwarf goats in cities comes from large dogs. It's very important that the area where the goats are kept—and preferably your entire yard—be dog proof.

Injury and illness: Be sure to periodically inspect your fencing and housing to ensure no sharp exposed edges, poking wires, or holes for escape exist. It's critically important that you keep the goat's grain secure and out of reach of the goats. If they get into the grain, they will overeat and experience bloat, which is potentially fatal. A goat who's experiencing bloat can be treated with Milk of Magnesia. Exercise caution in allowing your goats to have contact with other goats.

Be aware of the signs of goat illness, such as crying, decreased appetite, diarrhea, impaired gait, and listlessness. Keep an animal first-aid kit (see Chapter 10), use the goat message board if you suspect something may be wrong (see Appendix B), and keep the phone number of your livestock veterinarian handy.

Breeding and Kidding

Urban homesteaders approach breeding differently from their rural counterparts in a couple of ways. With limited space, you're likely not looking to breeding as a way to expand your herd. Rather, breeding is a means to an end. If you're raising dairy goats, periodic breeding is necessary to keep the milk flowing. Also, because urban homesteaders can't keep a buck on their premises, arranging to bring your doe to be "serviced" (really, that's what it's called!) by someone else's buck is part of the process.

Preparation, Breeding, and Pregnancy

As I previously mentioned, does can be bred when they're as young as 7 months old, but it's much better to wait until they're a year old before breeding. When you're ready to start the breeding process, find a breeder who has a buck who'd be a good match for your doe. The buck must be a miniature goat; otherwise, your doe could experience severe complications during labor.

The breeder will charge a "stud fee" and may have you sign a stud service contract. The stud fee could range from $25 to $50 for an unregistered buck up to $150 to $200 for a premium-quality registered buck.

Many doe owners arrange for what's called a driveway breeding, which means you bring your doe by to be serviced when you know she's in heat (ovulating) and take her home right afterward. Some breeders may offer the option of allowing your doe to stay with their herd for a little while so the doe is sure to be serviced when she's in heat.

Knowing when your doe is in heat can be tricky, but there are clear signs you can look for. Both Nigerian Dwarf and African Pygmy goats can be bred year-round, so you can begin looking for signs at any time. The signs of heat include …

- Flagging (rapidly waving the tail back and forth).
- Swollen and/or pink vulva.

- Discharge from the vulva, which sometimes causes the tail to become dirty.

- Increased bleating.

SMALL STEPS

If you want extra help determining when your doe is in heat, ask your breeder for a buck rag. The breeder will take an old rag, rub it all over one of his or her bucks, and put it in a sealed jar for you to take home. When you suspect your doe is in heat, open the jar under her nose. If she becomes excited, she is ready to be bred!

Does cycle approximately every three weeks, so it's helpful to begin charting the cycle ahead of when you'd like to attempt breeding. Does are generally in standing heat (which means they'll allow a buck to mate with them) for just 12 to 48 hours, so once you notice the onset of heat symptoms, you'll need to get the doe to the breeder's quickly. Your breeder can walk you through the specifics of the actual breeding, but it's pretty straight-forward. You usually don't need to worry about your sweet little doe getting mauled by a big buck—most of the time, the female goat totally runs the show. Your breeder will be able to tell you when the goats have succeeded in mating and when they're done.

Does have a gestation period of approximately 150 days. They typically carry two kids, although anything from one to four kids is possible. It's often difficult to tell if a doe is pregnant for the first three months or so. However, if you're used to handling your doe regularly, you'll likely notice changes in her girth before then. Also, you can watch for signs of heat around three weeks after the breeding. If she goes into heat, you should try the breeding again. If not, she's likely pregnant.

ROAD BLOCK

Although Nigerian Dwarf and African Pygmy goats can be bred at any time of year, some times are better than others. Take care to schedule the breeding so the kids won't be born during the coldest months of the year.

Does need a little special care during their pregnancy. It's best to feed alfalfa hay during this time, and be sure she gets some grain mixture twice a day. If the grain does not already contain molasses, you will want to add some to the grain—particularly during the last month of her pregnancy. If you're not already giving your goats mineral salts, you should make them available during pregnancy.

Check with your veterinarian to see if he or she recommends any vaccinations for your doe during pregnancy. Sometimes does are given a shot at four months gestation to effectively vaccinate both them and their fetuses.

Kidding

The most important thing to remember is that the vast majority of goat kiddings are uneventful. Does are well equipped to birth their kids without assistance, and your role will be minimal. If the doe is having difficulty, there are some things you can do to help, and you'll want to have your veterinarian's phone number ready in case serious problems arise. Talk with your livestock vet early on to be sure he or she can make house calls in situations like this. If you'd like to familiarize yourself with the visuals of a goat birth ahead of time, the Filaree Farm website offers a great photo slideshow (see Appendix B for the direct link).

ROAD BLOCK

Even the most supportive neighbor is likely to be alarmed at the noises your doe will make when she's in labor. Take some time before the kidding to talk with your neighbors about what will happen. Be sure to let them know that the kids will be moving on to a new home once they're weaned.

In addition to the standard animal first-aid kit (see Chapter 10), here are the supplies you'll want to have on hand prior to kidding:

- Towels

- Hair dryer (if the weather is cold)

- Scissors and string

- Bulb syringe

- Flashlight

- Goat vitamin mixture with A, D, and E (Nutri-Drench)

- Molasses

- Powdered *colostrum*

DEFINITION

Colostrum is a thick milk the doe produces immediately after giving birth. It's rich in nutrients and antibodies, and without it, a newborn kid has a low chance of survival.

A normal kidding will have four stages:

Stage 1: The doe will exhibit "nesting" behavior, and she may be less likely to get up and move around the barnyard. She may be shifting uncomfortably. There will be some jelly-like discharge from her vulva. This stage can last a full day.

Stage 2: The birth membrane protrudes (it will look like a pink bubble coming out of the vulva), and the water will break. You can expect delivery to begin 30 minutes after the water breaks.

Stage 3: The doe will deliver the babies. The first kid will be more difficult to push out, but subsequent kids will likely come quickly.

Stage 4: The doe will expel the placenta. This could be as long as three or four hours after the birth.

If your goat has been in labor for an extended period of time without producing a kid, or if you've attempted to assist her and the kid seems to be stuck, it's probably time to call for help. The Goat Spot forum (see Appendix B) has a list of emergency numbers you can call for advice during a kidding, and if necessary, you should be prepared to call your veterinarian for a home visit.

It's a good idea to learn more about kidding prior to your doe's due date—the goat books and websites listed in Appendix B provide additional information.

After the delivery, you can help the doe dry off her kids using the towels (or, if necessary, a hair dryer). Don't dry the kids completely because this will remove their scent and the mother may reject them. Use the bulb syringe to clean out the kids' nose and mouth. Remove any wet bedding from the area around the doe and kids.

Give the doe some water with a little molasses added and give the kids the vitamin mixture using a needleless syringe. (Follow the bottle's instructions regarding the amount to give.) Tie each kid's umbilical cord with string and cut it with scissors that have been disinfected. Mix a solution of iodine and water, and use it to clean the umbilical stub. (This will need to repeated daily for a week.)

Watch the kids carefully to be sure they're successfully nursing. It can take kids up to an hour or two to get the nursing thing figured out. However, if they're too weak to nurse or the mother won't allow it, you will need to use a needleless syringe to give the kids colostrum.

Raising and Selling the Kids

Watching baby goat kids frolic in their yard is one of the great joys of urban homestead-ing. The situation is made even better because you don't have a lot of responsibilities in raising the kids. Their mother, and any other goats in the pen, will take care of that. Just provide the mother with good food, be sure the pen is safe and secure, and everyone will be fine. The kids will nurse exclusively for the first few weeks and then they'll gradually start to eat more hay and less milk.

Sometimes a doe will reject a kid, and if that's the case, you need to milk the mother and bottle-feed the kid. But this is a rare occurrence.

Talk with your veterinarian about the recommended vaccinations for the kids. Often they need to be vaccinated at 4 weeks and then receive a booster at 8 weeks.

Many goat owners—both in the city and in rural areas—choose to *disbud* their goats. This procedure happens when the kids are around 3 weeks old and can be done by your livestock veterinarian. Disbudding is painful for the kids for the short duration of the procedure, but it's beneficial for their health and safety in the long run. Goats with horns can seriously injure each other—as well as their human caregivers—during the course of normal goat play. Plus, goats can sometimes get their horns caught in fencing, which can cause damage to them as well as the fence.

DEFINITION

Disbudding is a procedure that prevents the growth of horns.

The kids should stay with their mother for eight weeks. After that, they're ready to be weaned and moved to their new home. Once the kids are born, you should begin advertis-ing them for sale. This can be done through your local goat associations, 4-H clubs, homesteading groups, and on Craigslist. It's good practice to complete the disbudding and all recommended vaccinations before handing the kids off to their new owner.

If you have a male kid, you need to decide whether you're going to try to sell him as a buck or a wether. If you'd like to make him a wether, your veterinarian can perform the castration procedure. This typically happens around 3 weeks of age, but talk to your veterinarian to get his recommendation and more information. Be aware that male kids become sexually mature at 7 weeks, so you'll need to wether or rehome him by that time!

Milk, Fiber, and Meat

Goats can be so much fun to have around, it's sometimes easy to forget that they also provide food or fiber for you and your family. Goat milk is consumed by more people around the world than cow milk, and goat meat is commonly used in Middle Eastern, Indian, Pakistani, Mexican, and Caribbean cuisine. The mohair produced by fiber goats is highly prized by textile producers.

Milking Basics

Learning to milk is a bit like learning to drive a car. The process takes time and seems complicated at first, but once you get the hang of it, the task becomes automatic and you can't remember what all the fuss was about.

You can read about how to prepare supplies for milking and take care of the milk afterward, but when it comes to actually getting the milk out of the goat, there's no substitute for watching someone do it—and then practicing on your own goats. Your breeder will probably show you how to milk if you ask, or you can contact a local goat owner and offer to spend a couple hours cleaning his or her barnyard in exchange for a milking lesson.

You can start milking your doe a little while she's still nursing her kids. You won't get much milk, but starting early gives both of you time to get used to milking before the kids leave and the milking becomes mandatory. Goats need to be milked regularly in order to keep the milk flowing. Otherwise, the doe will "dry up" early. The standard practice is to milk twice a day at approximately 12-hour intervals. However, you can milk your goats just once a day if you'd like—you just won't get as much milk. A natural decline in milk production over the goat's lactation period is to be expected. The more closely you hold to an every-12-hour milking schedule, the more milk you can expect from your goat.

You may be able to find milking supplies at your feed store, but it's likely you'll have to order them online from a goat supply company. Here are the supplies you need to milk your goats:

- Milking stand*
- Seamless stainless-steel bucket (2 quart is a good size for miniature goats) or a 1 quart Pyrex measuring cup
- *Strip cup*
- Udder wash—1 quart water, 1 teaspoon bleach, a few drops liquid dish soap (Bleach can be omitted for goats with sensitive skin.)

- Teat dip (available commercially) or an iodine or nolvasan/chlorhexidine solution

- Small container with lid (like Tupperware) for teat dip

- Stainless-steel milk strainer

- Filters

- Glass jars (for storing the milk)

- Clean rags

- Paper towels and small binder clips (optional)

Milking stands elevate and secure the goat, making milking easier. See Appendix B for a website with milking stand building plans.

DEFINITION

A **strip cup** is a cup with a wire mesh filter. The first squirts of milk are directed into the strip cup when milking, and the filter will show any abnormalities that may be present in the milk. In addition, the first few squirts of milk contain the most bacteria, so it's best to discard them.

Here are the basic steps to follow when milking your goat:

1. Sterilize the pail (or Pyrex cup) and strainer in a bleach water solution.

2. If you'd like to prevent hair and debris from falling into your milk, you can clip a clean paper towel over the opening of your pail. (The clips need to be sterilized in the bleach water solution.)

3. Latch your goat into the milking stand and use a fresh rag to thoroughly clean her udder with the udder wash solution.

4. Wash and dry your hands. Milk the first four squirts from each teat into the strip cup.

5. Begin milking into your pail and continue until it seems you can't get any more milk out of the udder. At this point, you can "bump" (similar to what kids do when they're nursing) or massage the udder to stimulate a little additional flow.

6. When you've finished milking, briefly dip each teat into the teat dip cup to help prevent mastitis.

Be sure the goat's water trough is full because she'll likely be thirsty after milking.

SMALL STEPS

Most goat owners only give their lactating goats grain when they're on the milking stand. It's a great incentive to get your doe to hop onto the stand on her own, and eating the grain keeps her occupied during milking.

It's important to cool the milk right after you've completed milking. Place the strainer and filter on top of your glass jar, pour in the milk, cover, and refrigerate. It's a good idea to date your milk. Fresh goat milk is mild and sweet, but it does take on more of a "goaty" flavor after three days and is better used for cheese or yogurt before that point.

If you want to pasteurize your milk, you can heat it in a double boiler at 145°F for 30 minutes or 163°F for 30 seconds before cooling. The decision of whether to pasteurize your milk or leave it raw is a personal one, and you should make the choice best for you and your family.

Collecting Fiber

The first step in getting fiber is to know what kind of fleece your goat has because that will affect the way you collect it:

- Type A goats have long, silky hair that hangs in curly locks about 6 inches long. Type A goats must be shorn.

- Type B goats are the most common and have shorter hair, about 3 to 6 inches long, with a nice crimp. Type B goats can be shorn, combed, or plucked.

- Type C goats have a coarse-looking coat, but the hair is actually a very fine fiber that's about 1 to 3 inches long. They can be shorn, combed, or plucked. If you have a Type C goat, you can even let the coat blow off in the wind and then collect the fibers.

The best time to harvest fiber from your goats is in the late winter before the start of spring. Be sure to supply adequate heat and bedding for your newly "naked" goats after you take the fiber.

To shear a goat, use an electric shear (with a goat comb) or a pair of sharp pet shears. If you'd like to comb or pluck your goat, you'll need to brush her periodically to discover when she's ready to start shedding her fleece. You can use a plastic hairbrush, a pet grooming brush, or a cotton carder. Goats don't shed their entire coat at once, so you will need to comb your goat every few days. The hair should release easily—if it doesn't, wait a couple of days before combing or plucking again.

Harvesting Meat Goats

Goat meat is consumed all over the world. It's sometimes called *chevon*, and the meat from kids is also known as *cabrito*. Wether goats are often harvested for meat, while the does are kept for breeding and milking.

I don't recommend slaughtering your goats on your property. You can search online to find a farmer who raises meat goats and talk with him about bringing your goats to him to be harvested. Halal and kosher butchers often know how to slaughter goats. If they're not able to do it for you, they can likely direct you to someone who will. If there are any small slaughterhouses in your area, they might be able to butcher the goat for you. The shops that process deer carcasses for hunters might also be able to process your goat into cuts of meat for you.

The Least You Need to Know

- Miniature goats can be comfortably raised in an urban backyard.
- Goats are not difficult to care for as long as they're provided with food, water, shelter, and a secure yard.
- Dairy goats must be bred periodically in order to produce milk, and lactating goats need to be milked twice a day.
- Miniature goats can also be raised for fiber (mohair) and meat.

Raising Rabbits

In This Chapter

- Selecting rabbits for meat or fiber
- Caring for your rabbits
- Successful breeding
- Harvesting meat and fiber

Rabbits have been called the ideal urban livestock. Whenever space is at a premium, rabbits are especially valuable. They're clean, quiet, and can be kept just about anywhere. Rabbits are inexpensive to purchase and raise. A small trio of rabbits can produce an impressive 150 pounds of lean, healthy meat per year. Plus, rabbits also create manure that's the perfect fertilizer for urban gardens. It's also possible to raise rabbits for fiber and spin the wool to create beautiful (and warm) clothing.

In this chapter, I talk about how to select your rabbits, whether you're interested in raising them for meat or fiber. I cover the basics of caring for your rabbits and tell you how to provide housing that will keep them safe and secure. I discuss breeding rabbits and explain how you can support the mother in raising the babies. I also go over harvesting the meat and fiber from your rabbits.

Choosing Your Rabbits

Most urban homesteaders who raise rabbits do so for meat production. Pelts are also a by-product of meat rabbits, although the work of curing and drying the skins can be considerable. Some rabbits can be raised for fiber, but it may take a while to collect enough from a single rabbit to make a pair of mittens! However, all rabbits do provide wonderful manure and can be worth raising for that alone.

ROAD BLOCK

Despite what you may have seen magicians do, you should never pick up a rabbit by the ears. You can lift rabbits by the loose flesh at the back of their neck, but always support their bottom at the same time. If you can, carry them with their head tucked beneath your arm—it makes them feel secure.

Rabbits for Meat

Although there are around 50 breeds of rabbits, homesteaders prefer a handful for the purpose of raising meat. New Zealand Whites and Californians are two of the favorites. These are moderate-size rabbits, which means they're around 5 pounds when they're ready for harvesting (at 8 weeks old) or about 10 pounds when fully mature.

Both New Zealanders and Californians are known for their efficient growth, disease resistance, good bone-to-meat ratio, fertility, and large litter sizes. Plus, both breeds have white pelts, which are often preferred because they're easier to dye.

A couple of smaller breeds also work well as meat rabbits and can be raised in less space than a New Zealand or Californian. Tans and Florida Whites take longer to reach a good size for harvesting (10 to 12 weeks old instead of 8), but they have finer bones and therefore produce more meat per pound. Tans and Florida Whites have smaller litters than larger breeds.

Some large rabbit breeds, like the Flemish Giant, can grow to 14 pounds or more. These breeds are not the best for meat production because they consume disproportionately more feed for their size and therefore are not efficient feed converters.

Rabbits have a reputation for being prolific breeders. This reputation has been well earned, and it's a gift to an urban homesteader. Two healthy does and one buck are sufficient for beginning and maintaining a respectable home meat rabbit operation. One doe can produce around 75 pounds of meat per year. With two breeding does, you can have a rabbit dinner at least once a week.

Although rabbits can technically become pregnant when they're just 8 weeks old, waiting until they're around 6 months old is best. It also becomes easier to accurately determine the gender of rabbits as they become older, and you don't want to accidentally purchase a doe in place of a buck or vice versa. So beginning with a trio of 6-month-old rabbits is a great way to start. If you'd like to get a head start, you can try to find pregnant does and purchase them along with the buck they were bred to.

Rabbits for Fiber

Angora rabbits can be raised for fiber. The wool they produce is silky and can be used for yarn and fabrics. It is high quality and valuable, but an angora rabbit will produce only about 2 or 3 ounces of fiber every three months.

Even if you don't plan on using the wool, angoras must be brushed regularly to keep them clean and healthy. Although all rabbits dislike hot weather, it's especially dangerous for angoras. Also, extra care needs to be taken to keep their cages clean because the hair can build up on the cage floor and trap droppings.

How to Purchase

A good place to begin your search for rabbits is by visiting a feed store. It will likely have a bulletin board where you can find information about rabbits for sale in your area, and the feed store clerks may also be able to point you toward customers who breed rabbits. You also can contact the local 4-H club or attend a rabbit show. (Visit the American Rabbit Breeders Association [ARBA] website at arba.net to find a show near you.) Some breeders registered with the ARBA will even agree to ship rabbits to you, and their rabbits are typically very high quality.

It's a good idea to visit several *rabbitries* before making a purchase so you can get a clear idea of what's available. It's not wise to cut corners when selecting your breeding rabbits, considering they're the basis of your whole operation. A good rabbit is clean and alert. Don't get a rabbit who's droopy or listless. Rabbits are nocturnal, so a little bit of daytime sleepiness is to be expected, but handling the rabbit should show you if she's alert or not. Rabbits should be neither too fat nor too thin. Examine the rabbit's ears, eyes, teeth, and nose for any signs of disease. Check to be sure there's no red or irritated skin around the rabbit's genitals (it would look like diaper rash), which is caused by ammonia from soiled bedding. Any rabbit living in those conditions is likely to have additional problems.

DEFINITION

A **rabbitry** is a place where domesticated rabbits are kept and typically bred.

After doing research and determining which breed is best for you, you'll want to be sure the rabbits you're getting will perform the way they're supposed to. A white rabbit may look like a New Zealand White but could end up being a different breed (or combination of breeds) that doesn't have the meat production qualities you're looking for.

It's advisable to spend a little extra money on rabbits who are pedigreed purebred, meaning their lineage is documented. Rabbits that are not pedigreed are considered to be crossbred, even though they may actually be purebred (but there's no way to prove it). Some rabbits are also registered, which means they've been examined by an official breed registrar and meet the breed standard. While these rabbits are clearly high quality, buying a registered rabbit would be more important for someone interested in entering rabbit shows or breeding rabbits for sale (as opposed to home meat or fiber production).

You can expect to pay anywhere from $15 to $40 for a good meat rabbit and around $50 to $60 for an Angora.

Rabbit Care

Rabbits are lauded as being easy to care for and relatively trouble free, at least when compared to other livestock. That being said, there's a difference between tending to one rabbit as a pet and entering the realm of raising multiple rabbits. You will need to plan ahead and be sure you have adequate housing for your rabbits and their offspring, good equipment for feeding and watering, and methods for protecting your rabbits from predators and extreme temperatures. If you create the right conditions for your rabbits, they should stay happy and healthy.

SMALL STEPS

Rabbits have a reputation as the easiest food-producing animal to care for. They should be given food and fresh water each day, which takes just a couple of minutes per rabbit, and you'll spend a little time once a week to clean.

Housing

Good rabbit housing needs to provide the following elements:

- Adequate space
- Ease in keeping clean
- Good ventilation but protected from drafts
- Shelter from temperature extremes and precipitation
- Protection from predators

Although there are many ways to keep rabbits, it's generally agreed that all-wire cages are the best way to achieve the necessary shelter goals. Any sort of hutch that uses wood as a building material is asking for trouble. The rabbit will gnaw the wood and soak it with urine and then it will rot. Any time the droppings are not able to fall through wire mesh (for example, in the corner of a wood frame), they will pile up and become difficult to clean.

It's best not to build cages out of chicken wire because it's easy for predators to tear through. The recommended material is 14-gauge galvanized wire mesh.

SMALL STEPS

If you construct a cage with wire mesh, be sure the smooth side is facing up when you construct the floor. This is much easier on the rabbit's feet and will prevent problems with sore hocks (the first joint, behind the foot) down the road.

The floor should be $\frac{1}{2}$×1 inch, and the walls and ceiling can be 1×2 inches. A cage that is 36×36 inches and 18 inches high will do well for a doe and her babies.

Breeding does should not share a cage (although it's okay to put two does in a cage before they're bred as long as they get along), and your buck will need his own cage, so if you're starting with a trio of rabbits, you should plan on three cages. (See Appendix B for a resource with detailed instructions on building cages.)

Rabbits need to be sheltered from heat and precipitation, so the cages should be inside some sort of shelter. It's entirely possible to raise rabbits inside. Many wire cages come with a pull-out tray (which should be filled with straw or pine shavings) under the floor to catch droppings and urine, so the cages can be stacked if needed to save space. If you'd like to keep the cages outdoors, you can put them in a ventilated garage or shed.

Many experienced rabbit raisers prefer to suspend their wire cages from the ceiling (at a convenient height for the caregiver) rather than placing the cages on supports or on the ground. This makes it more difficult for predators and rodents to access the cages, and it prevents the aforementioned rotting that urine would cause to wood supports. If the cages don't have a pull-out tray beneath the floor, they should be suspended over straw so it will catch the droppings and urine.

Raising rabbits in cages may be unattractive to homesteaders, who are generally looking for "free-range" alternatives for their meat. Cage advocates say cages allow for selective breeding and better breeding outcomes, better predator protection, and no injuries due to rabbit fights.

Other owners raise rabbits in open runs rather than cages. A run can be a secure building (like a shed) with good ventilation. It must be protected from digging—both the rabbits digging out and predators digging in—by securing wire mesh over the entire floor and then covering it with a thick layer of straw bedding. It's also possible to construct a moveable rabbit run (similar to a chicken run—see Chapter 11), as long as it has an attached shelter so the rabbits can get out of the sun or rain. It is not advisable to let your buck live in the run with the does.

Some people allow their rabbits to run loose in their homes. It isn't difficult to litter box–train a rabbit. They prefer to always go to the bathroom in the same place, so if you place some droppings in the litter box, they should begin using it from that point on. If you do let your rabbits roam in your home, you'll need to "rabbit-proof" your wiring with hard plastic tubing. Be aware that rabbits like to chew, so furniture legs are also at risk.

Raising Rabbits Without Land

Rabbits are so commonly raised in wire cages, as previously mentioned, that it's completely feasible to keep them inside your home. Many rabbit owners like to stack their cages, so you can keep several rabbits with very little floor space.

Putting your rabbit cages on your balcony leaves them vulnerable to the elements, but you can do it if the balcony is on the north side of the building—and therefore won't get too hot—and if you hang tarps to protect the rabbits from the elements.

If you want, you can use some of the ideas discussed in Chapters 11 and 12 to raise your rabbits on someone else's land, but it would be much simpler to just keep your rabbits indoors.

Food and Water

Rabbits are herbivores, but that doesn't mean they can eat any and all kinds of vegetables. Rabbits have sensitive systems (more so than chickens or goats) and should be fed with care. They need to have feed that provides them with roughage, protein, vitamins, and minerals. The best way to ensure a balanced diet for your rabbits is to feed them pellets that have been specially formulated for them.

Some people do make their own rabbit feed from scratch or forage/grow items to give to the rabbits. It may seem more economical than purchasing pellets at first, but over the long term, the rabbits will be much healthier due to the balanced nutrition from the pellets, and that will end up saving you money.

Rabbits also enjoy a mineral salt block in their hutch. This not only provides them with extra nutrients they need, but it gives them something to chew on.

It's okay to occasionally give rabbits "treats" from your garden or table, but be sure to introduce any new foods slowly. Rabbits can have small amounts of the following items:

- Corn (and corn stalks)
- Sunflower and pumpkin seeds
- Roots and greens of carrots, turnips, and beets
- Dark leafy greens
- Dried bean or pea vines
- Bread
- Apple and banana peels
- Alfalfa hay
- Dandelions and clover (except sweet clover)

ROAD BLOCK

Rabbits should not be fed any greens that have been sprayed with pesticide. Also, cabbage should not be given to rabbits.

The best way to give rabbits their pellets is to purchase a feeder that can be attached to the wire cage. Many of these feeders can be filled from the outside (without opening the cage), which makes feeding time easier for you. You can place the rabbit food in a bowl if you'd prefer, but rabbits tend to turn over bowls and waste food when it's given to them that way.

An adult rabbit who's not pregnant or lactating should get around 6 ounces of pellets each day. Pregnant or nursing does, and growing youngsters, should receive more than that. You'll get a sense for how much to feed your rabbits with experience. They should be fed once a day on a consistent schedule. The rabbits should be excited (but not frantic) to see you at feeding time, and they should eat all the pellets you give them. If food is regularly left over the following day, you're feeding too much.

You can use the time when your rabbits are eating to run your hand over their body and see if they feel too fat or thin. Overweight rabbits will not breed well, so try not to overfeed.

A regular supply of clean water is vital for keeping your rabbits healthy. A doe and her *litter* can drink up to a gallon of water on a hot day. Providing water to your rabbits in a dish will likely result in fouled water—plus, the rabbit will probably flip over the dish. It's best to buy a ballpoint waterer similar to what's used for hamsters and guinea pigs. You can find these at a feed store, pet store, or online.

DEFINITION

A **litter** is the group of rabbit babies produced in a single birth.

If you're raising your rabbits outdoors, you need to take steps to be sure the water doesn't freeze during winter. You can get multiple waterers and rotate them (replacing a frozen one with a fresh one a couple of times each day), or you can buy a heated waterer from the feed store. These have to be plugged in, but they prevent the water from freezing. Make sure that your rabbits can't chew through the electrical cords! Be sure to clean and sterilize your waterers at least once a week.

Maintenance and Challenges

Rabbits don't need much ongoing maintenance other than a clean living environment and good food and water. You may find that you want to occasionally clip your rabbit's nails to make them easier to handle; dog nail clippers work well for this.

Here are a few additional issues to keep in mind:

Extreme temperatures: Rabbits don't do well in the heat—in fact, the ideal temperature for them is around 50°F. They enjoy a little sunshine as long as they don't get too warm, but they definitely need to be shaded during the hotter parts of the day. Be sure to provide plenty of water during warm weather. You can also fill plastic jugs with water, freeze them, and put them in the cages to cool things down. Rabbits are better equipped for cold weather, but it's a good idea to have a lightbulb in the cages to provide supplemental heat if needed.

Teeth: Rabbits love to chew, and if you oblige that tendency with a chew toy, they will be more content. You can create a combination toy and salt lick by soaking a piece of 2×4 (non–chemically treated wood only) in very salty water for three days, air-dry it, and then give it to your rabbit.

Some rabbits have buck teeth (extra-long upper incisors) or wolf teeth (extra-long lower incisors). This can affect the rabbit's ability to eat normally, and the long teeth could eventually pierce the flesh of the mouth. This trait is hereditary, and affected rabbits should not be bred.

Predators: Rabbits are especially vulnerable to—and attractive to—predators. In addition to urban predators like foxes, raccoons, and dogs, even smaller animals like cats and rats can kill a rabbit. While using a deterrent like coyote urine (available online or at a sportman's store) can be helpful, the best way to protect your rabbits is to be sure their housing is secure.

Wire cages should be well made and suspended from the ceiling if possible (instead of positioned on supports or the floor). The shed or garage that houses the cages should be well ventilated but also predator proof. If the rabbits are not in cages, the floor needs to be covered with wire mesh to prevent digging in or out. (See the "Housing" section earlier in this chapter.)

Injury and illness: Be sure the rabbit cages are sheltered from above to prevent wild birds from defecating in the cages and giving the rabbits coccidiosis.

Rabbits raised on wood or other solid floors can get hutch burn from the excess ammonia in urine-soaked bedding. Hutch burn is red, irritated skin around the rabbit's genitals. In should be treated by removing the dirty bedding, and you can smear petroleum jelly over the affected area until it heals.

Rabbits sometimes get conjunctivitis (just like pinkeye in humans), which can be treated with an antibiotic eye ointment from the veterinarian.

If you notice a brown discharge or crust on your rabbit's ears, she may have ear mites. An infected rabbit should be isolated, and the mites can be treated with medication. Another option is to put a few drops of olive oil into the ear (you can also coat the inside of the ear using a cotton ball) and gently massage the base of the ear. Repeat every other day for two weeks, then once a week for two more weeks.

Rabbits sometimes contract a disease called Pasteurellosis (also called "the snuffles"). It is characterized by excessive sneezing and a snotty nose with white discharge. Does can infect their kits, and infected rabbits can contaminate others. You can put a partition between the cages to reduce this possibility, but isolation of the infected rabbit is recommended. Pasteurellosis must be treated with antibiotics.

It's a good idea to keep an animal first-aid kit on hand (see Chapter 10) in case of emergencies.

Creating More Rabbits

Healthy rabbits are typically cooperative breeders, and does are able to successfully raise their litters with very little help from you. However, it's important that you create an environment that supports your rabbits during breeding, birth, and the raising of the babies. Trying to breed rabbits in an improper setting could result in harm to your rabbits and could possibly cause the mother to destroy her litter.

Breeding and Pregnancy

Rabbits are ready to breed when they're around 6 months old. Before breeding, be sure each doe has her own cage. To breed the rabbits, place the doe in the buck's cage. Do *not* put the buck in the doe's cage because the doe will defend her space and may injure the buck. Don't leave the rabbits alone during breeding.

The rabbits will mate once or twice and then you can return the doe to her cage. If the buck falls over, it's an indication of a successful breeding. There's no need to watch for signs that the doe is in heat before breeding. In rabbits, breeding induces ovulation. Because of this, some rabbit owners bring the doe back to the buck six hours later for a second breeding to ensure all the eggs she ovulated get fertilized.

A rabbit's gestation is around 31 days. Pregnant does should receive a feed with at least 16 percent protein, and during the last half of the pregnancy, you can let her eat all she would like.

Does need to feel secure and sheltered when they're caring for their kits. If they feel stressed or threatened, they may destroy the litter. About five days before *kindling*, give your rabbit a nesting box. This is a partial enclosure, about 12×24 inches, where the doe will have her *kits*. It has an opening for the doe to jump in and out, with a lip about 6 inches tall to prevent the kits from stumbling out. The box's solid floor also helps keep the kits in the cage—if they're born on the wire mesh floor, they may fall through the holes.

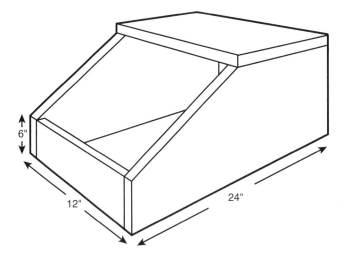

The nesting box allows the doe to feel sheltered and keeps the kits warm and secure.

DEFINITION

Kindling is the act of a rabbit giving birth. A **kit** is a baby rabbit.

You can construct a nesting box out of wood, although it will rot eventually. You can buy metal nesting boxes, which are easier to clean and more durable. If the metal nesting box is made of a fine mesh, place cardboard on the floor of the box during cold weather to block drafts.

Place a couple of inches of soft straw or pine shavings in the nesting box before putting it in the cage. The doe will spend a few hours prior to kindling getting the box ready. This includes pulling soft fur from her belly to line the nest.

Other than the nesting box, there isn't anything else you need to provide for the doe prior to kindling. The best thing you can do is leave her alone and give her peace and quiet.

Kindling

Your hope is that the doe will give birth to all her kits inside the nesting box. If some are left on the floor of the cage, they will quickly die of exposure unless they're moved into the nest. Temporarily remove the doe from the nest and place the babies inside. Before returning the doe to the nest, dab a little chest salve (like Vicks VapoRub) or vanilla extract on her nose. This will temporarily block her sense of smell and hopefully prevent her from rejecting the relocated babies.

SMALL STEPS

Try to handle the kits as little as possible. If the doe smells something foreign on the babies, she may reject the litter.

If all the babies were born in the nest, wait until the second day to check on them. Remove the doe, count the babies, and remove any that are dead. Use the chest salve (or vanilla extract) on the doe's nose, and place her back in the nest.

Does only have eight nipples, so if she has more than eight kits, you can try to foster the extras with another doe who has a smaller litter, provided both does kindled at the same time.

It's a good idea to keep breeding records for your does. Record the dates of kindling and the number of kits per litter.

Raising Kits

When the kits are 10 to 14 days old, they'll begin opening their eyes, and at 16 days, they'll start venturing outside the nest. They are still nursing during this time, so you don't need to provide any special food for the babies. At 21 days, they'll start eating solid food. You can provide them with a little bread soaked in milk, but be careful when introducing any other new foods because young rabbits have delicate digestive systems. They should not be given any greens of any kind until they're adults because greens can give them diarrhea.

The kits can stay in the cage with their mother until they're 8 weeks old. At that time, they'll be around 5 pounds and are ready to be harvested. If you're not going to harvest the kits yet, or if you'd like to save some of them as breeding stock, you need to remove them from the doe's cage and separate the males from the females. Although rabbits shouldn't be bred until they are 6 months old, it's possible for 8-week-olds to mate. Rabbits are almost impossible to sex until they're a couple of months old. At that point, the bucks' testicles will begin to protrude. Determining rabbit gender takes practice, but in general, the male sex organ looks more round and the female's looks more like a slit.

Eight-week-old rabbits who aren't going to be harvested can be kept in cages (two does to cage, but each buck needs his own cage) for another five weeks or so. After that point, the feed-to-meat conversion ratio drops sharply, and it's not economically practical to keep the rabbits longer than that unless you're going to use them for breeding.

Once the kits leave the doe she can be rebred, which would give you four litters per doe per year (one month gestation and two months raising per litter). Some rabbit owners rebreed their does as early as four or five weeks after kindling and remove the kits when they're 6 or 7 weeks old. This gives the doe about two weeks without babies to build her strength for the next kindling.

Meat and Fiber

Although it's certainly fine to keep a rabbit as a pet, homesteaders are typically interested in rabbits as a source of food or fiber. It might seem strange to raise meat animals in the city, but rabbits' small size and modest space requirements make it possible. However, I still recommend taking your rabbits out of the city before harvesting.

Harvesting Your Rabbits

Currently, rabbit meat is not widely consumed in the United States. But it is a common food in many parts of the world, and it's appearing more frequently on the menus of some of the finest restaurants. Some people have compared the taste of rabbit to chicken, but veteran rabbit raisers say it has a flavor all its own. Homesteaders have used rabbit in a number of recipes over the years, including fried rabbit and rabbit with dumplings. Rabbits have all white meat, and it's leaner than any other land meat, as the following table illustrates.

Nutritional Values of Common Meats

Meat (Edible Portion, Uncooked)	% Protein	% Fat	% Moisture	Calories per pound
Rabbit (fryer, with giblets)	20.8	10.2	67.9	795
Chicken (fryer, with giblets)	20.0	11.0	67.6	810
Veal (medium fat)	19.1	12.0	68.0	840
Turkey (medium fat, with giblets)	20.1	20.2	58.3	1,190
Lamb (medium fat)	15.7	27.7	55.8	1,420
Beef (fat)	16.3	28.0	55.0	1,440
Pork (medium fat)	11.9	45.0	42.0	2,050

Taken from Circular No. 547, U.S. Department of Agriculture, Washington, D.C.

While it can be challenging to slaughter an animal you've raised, rabbits are considerably easier to harvest than chickens—for starters, you don't have to pluck them! If you're thinking of raising meat rabbits, you should find a rabbit farmer in your area and offer to help with the harvest. This will give you a good idea of whether you're up for the challenge of killing and skinning your own animals. A number of books and videos detail how to process a rabbit. (See Appendix B for resources.) The whole process—from live rabbit to oven-ready meat—takes only a few minutes for those who are experienced with it.

SMALL STEPS

Pelts (the skin of an animal with the fur still attached) can be a by-product of meat rabbits. However, rabbits harvested when they're young are considered to have underdeveloped fur. In France, rabbits are kept longer before butchering so the fur is thicker.

Collecting Fiber

Angora rabbits produce beautiful fiber that's 4 to 8 inches in length. They molt about every three months, during which time the fiber will easily come loose with a gentle tug.

You'll know that the rabbit is beginning the molting period when you notice clumps of fur caught in the cage or hanging off the rabbit. Collect the fiber by combing it with a pronged comb or plucking it by hand. It can be spun with a drop spindle. (See Chapter 18 for more information.) Angora rabbits produce 2 or 3 ounces of premium fiber with every molt.

It's important to comb the rabbits regularly—at least once a week—to keep the fur from becoming matted or dirty between molting periods.

The Least You Need to Know

- Rabbits can be easily raised in an urban setting.
- Three rabbits can produce up to 150 pounds of healthy, lean meat per year.
- Breeding rabbits requires very little work, as long as the animals are given the proper environment.
- Angora rabbits produce beautiful fiber, which can be used for spinning.

Bee Busy

In This Chapter

- How the hive works
- What you need to begin
- Maintaining a healthy hive
- Harvesting honey and other treasures

Urban homesteaders can enjoy the sweet life by owning a beehive or two. Bees are silent, odorless (except for the sweet smell of the nectar), and use almost no space—all of which make them great candidates for the city. Not only do bees produce honey, but they're wonderfully helpful in the garden. Having bees nearby makes it more likely that your vegetable plants will be well pollinated, which leads to a better garden harvest.

In this chapter, I begin by discussing the basic workings of a beehive. I review the supplies you need to raise bees and weigh the pros and cons of two popular hive designs. I talk about how you can acquire your bees and what to do when you bring them home. I go over general hive maintenance and some of the challenges you may encounter. And of course, I talk about harvesting honey (and other goodies) from your hive.

Bee Basics

As a hive owner, your responsibilities are fairly limited. You can count on the bees to do the work of creating the honey, and they also take care of most of the hive maintenance. You just need to support their efforts and monitor things to be sure no problems arise. The inner workings of a healthy beehive are complex and interdependent. During the summer months, a beehive may have more than 60,000 bees, and each bee has a specific set of duties. It's a good idea to become familiar with the three types (or "castes," in bee lingo) of bees so you can understand some of what should be happening in the hive.

SMALL STEPS

It will take a little while to get your hive set up, and most new beekeepers spend extra time with their bees during the first year to learn the ropes. However, beekeeping is a fairly hands-off activity. On average, you can expect to spend 40 hours a year to maintain your hive and harvest the honey.

The Queen Bee

The queen is the heart of the hive. There's usually only one queen per bee colony, and most of what happens in the hive centers around her.

The queen has two main duties. The first is to lay all the eggs for the hive—she's the only reproductive female in the colony. The queen takes a series of mating flights in the spring, during which she mates with the male bees from the hive. During these mating flights, she stores up enough sperm to fertilize hundreds of thousands of eggs. She then stays in the hive to lay eggs—sometimes more than 1,500 a day!—until the fall. Not every egg the queen lays is fertilized, though. Fertilized eggs become worker (female) bees, and unfertilized eggs become drone (male) bees.

The queen also produces a complex mix of pheromones collectively called the "queen substances." They're her way of communicating with the other bees and regulating what happens in the colony. The pheromones also inhibit the development of the worker bees' ovaries, ensuring that the queen is the only laying female in the hive. The queen substances are passed through the hive by the bees as they share food. If the queen is removed from the hive, the bees will be aware of her absence within hours.

The queen is the largest bee in the hive, with a smooth, elongated abdomen that extends beyond her wings. Queens can live for five years, but they're rarely productive for more than two. As the queen ages, her egg production decreases along with the amount of pheromones she emits. Some hive owners replace their queen every fall to ensure consistent egg-laying and a vibrant hive. However, if left to their own devices, the hive will rear a new queen of their own once the old queen ceases being productive. See the "Hive Maintenance" section later in this chapter for more information.

Worker Bees

The worker bees are every bit as busy as their name implies. These nonreproductive females take care of all of the work, both inside and outside the hive. The first three weeks of a worker bee's life are spent tending to tasks within the hive, during which time she is referred to as a "house bee." House bees aren't responsible for every task at once, but they take on a succession of jobs as they mature.

House bees clean the hive, remove dead bees, groom and feed the queen, tend to the developing larvae, take nectar and pollen from foraging bees and store it in the hive, and fan the hive with their wings to regulate temperature and humidity. They also stand guard and protect the hive from intruders, like ants or bees from a different hive. House bees also produce beeswax, which they use to build wax comb. The comb consists of different compartments, called cells, that are used to store nectar (which becomes honey) and pollen. They are also where the queen lays her eggs.

After three weeks of completing tasks inside the hive, the worker bees are ready to begin foraging for nectar, pollen, *propolis,* and water. The worker bees, referred to as "field bees" during this stage, can venture as far as a 3-mile radius from the hive. Worker bees born during the spring or summer live for only six weeks. Those that reach maturity in the late fall can live as long as four months. They cluster around the queen bee and the brood to keep them warm during the winter, and they help raise the first batch of new bees in the spring.

DEFINITION

Propolis is a resinous substance bees collect from trees and plants. It's used as a sealant in the construction of their hives and also has antimicrobial properties.

Drone Bees

The male bees in the hive are drones. They don't take care of the hive, tend to the young, or forage for nectar. Their sole purpose is to mate with the queen. Drones follow her as she flies out of the hive, sometimes going a mile or more away from the hive and several hundred feet up in the air. The mating process is hard on the drones—they die after mating. Any drones left in the hive as winter approaches are kicked out by the worker bees, who don't want the drones consuming precious hive resources over the winter.

Drones are large, about twice as big as a worker bee, and are sometimes mistaken for the queen. However, their body is wide and the queen's is tapered. They also have large eyes that seem to cover their entire head. Drones don't have a stinger.

A hive has between several hundred and a thousand drone bees during the spring and summer.

The Supplies You Need

Beekeeping requires a modest initial investment. How much depends on which hive design you choose and whether you'll be building it yourself. Taking the time to acquire the proper equipment before getting your bees makes caring for them much easier.

Selecting a Hive Design

Two types of hives are most commonly used for home beekeeping: Langstroth and top-bar. Each has its advantages and challenges, but both are being used successfully in urban settings.

Langstroth hives are the type most commonly used by backyard beekeepers because they allow the beekeeper to check on the hive without damaging the comb. The basic structure of the hive is comprised of deep chambers. At the bottom is the brood chamber, where the queen lays eggs. Above that is the food chamber, where the bees store most of their honey and pollen. They're topped with "supers," which are removable compartments the worker bees use for honey storage.

Langstroth hive parts are widely available commercially, and the standardized sizing makes parts from different sources interchangeable. The internal parts of the hive are structured with precise $3/8$-inch spacings. This facilitates the building of even comb by the bees and prevents them from gluing open parts of the hive with propolis.

Langstroth hives cost around $150 to $200, but they result in the highest quantity of honey for the beekeeper. They're not easy to construct, but it can be done with detailed plans (see Appendix B). They're heavier than top-bar hives and more cumbersome to work with. Langstroth hives are more readily identifiable as beehives, which isn't always advantageous for urban beekeepers.

Top-bar hives are commonly used in developing countries and are also gaining popularity with urban beekeepers. They're easy to construct and can be built out of repurposed materials. A commonly used design looks somewhat like a feed trough, although almost any type of container can be used for the body of the hive. See Appendix B for a website that shows how to build a top-bar hive.

Bars are suspended across the top of the hive (hence the name), on which the bees can build their comb. Some beekeepers feel that this is a more natural way for bees to work because the structure of the comb isn't restricted, the way it is in a Langstroth hive. The natural comb design is also said to help the bees manage pests. However, since bees like to fill openings with propolis, the beekeeper often has to crack open the seal between the comb and the side of the hive to inspect the bees and harvest honey. The combs cannot be returned to the hive after harvesting, so top-bar hives yield less honey than a Langstroth hive.

A Langstroth hive with two chambers and one super (left) and a top-bar hive made from repurposed materials (right).
(Courtesy of BrianKraft.com)

Beekeepers are nothing if not an ingenious bunch and are constantly exploring different practices. For example, some have started using Warre hives, which are stackable top-bar hives that have some of the qualities of Langstroth hives. It's always a good idea to try to join a beekeeping club in your city so you can learn from others and stay up-to-date on the latest beekeeping developments!

Helpful Equipment

You'll need a few tools to effectively maintain your hive. Your best bet may be to purchase a kit that contains all these tools, which will cost around $95—and is cheaper than purchasing everything individually. You can also try to find used equipment through a local beekeeping club or bee supply store.

Here's what you'll need:

Smoker: The smoke will calm your bees and allow you to safely check on the hive and harvest the honey. Stainless-steel smokers are the most durable.

ROAD BLOCK

Some beekeepers believe the use of smoke creates stress for the bees, which can lead to hive health problems.

Hive tool: This flat, usually steel tool has many uses. It has a flat end for prying open the hive and a curved end for scraping and cleaning hive frames or the hive itself.

Veil: Although it may seem extreme to put on a veil, it's helpful to wear one each time you visit your hive. Honeybees are generally docile, but they do love to explore. They especially like dark holes … like your nose and ears—yikes! It makes sense to take reasonable steps to protect yourself.

Gloves: Gloves aren't always advantageous. They can make you clumsier and more likely to disturb or injure your bees. So it's best not to wear them except when necessary. This includes later in the summer when the hive is defending its honey, while you're harvesting, and when you're moving hive frames.

When working with your hive, you should always wear clothing that protects you. Long-sleeved shirts are good, as are pants you can tuck into your socks or boots. You can also use Velcro straps, similar to what bicyclists use, to secure your pant legs tightly around your ankles.

Be sure your clothes are clean because bees react to strong body odor. It's important to wash your bee clothes if you've been previously stung while wearing them because the old sting leaves behind a scent.

Acquiring Your Bees

You have a few options when it comes to obtaining your bees. The first is to order bees online. You can get "package bees" that are sold by the pound along with one queen. An order of 3 pounds will contain around 11,000 bees and is the right amount to get you started with one hive. You'd pay around $80 for this many bees. The bees will be delivered to your post office—which will surely call you right away and ask you to pick up the buzzing box!

Several different types of honeybees are available online, but Italian or Russian honeybees are the best choice for beginners. Both are docile and productive and can do well in a range of climates. Arrange to receive your bees in the early spring so they'll have time to establish their colony before flowers and plants start blooming. Plan to place your order early, in the late fall or winter, because some suppliers sell out.

SMALL STEPS

Beginning with one hive is a great idea for a new beekeeper. It's always good to start small and build from there. However, some novice beekeepers choose to get two hives so they can compare one to the other and get an idea of what is normal bee behavior.

You can also arrange to purchase bees from a local beekeeper. Often called a "nuc" (nucleus) colony, this will also contain thousands of bees and an actively laying queen.

If you'd like an inexpensive option—that also has an element of adventure—you can arrange to receive a swarm that's been captured by an experienced beekeeper. Bees leave their hive and form a swarm, which can be relocated to a new hive. Your local beekeepers association likely has a way for you to sign up to receive a swarm once it's been captured. Swarms should contain a queen, but the swarm catcher will have to take care to be sure that the queen makes it into the box during capture.

Getting Started with Your Hive

It's important to have everything set up and all your hive tools at the ready before your bees arrive. The trip to your home will be stressful for your bees, and you'll want to get them safely put into their hive as soon as possible.

Positioning Your Hive

Part of what makes beekeeping so appropriate for urban settings is that the hives don't take up much space. Some cities have rules about where hives can be located on your property, so check your zoning regulations before selecting a spot for yours. Here are a few additional things to keep in mind:

Hive entrance: Bees enter and exit the hive in a line, directly out the hive opening. Be sure this "bee line" doesn't cross a path frequented by people or pets. If you place the hive so the entrance is near a fence or wall, the bees will soar upward when they exit and do their initial flying above head-level.

Direction: If you face your hive toward the southeast, the bees will arise early and get a head start on the day.

Sunlight: It's not good to put the hive in full sun because the increased heat forces the bees to work harder cooling things down. However, it should not be located in complete shade, either, because that can lead to dampness in the hive. The ideal condition is partial, or dappled, sunlight.

Water source: Bees don't just forage for honey and nectar—they need water, too! Although the worker bees will range far and wide, providing a nearby source of water keeps them from focusing on something else, like your neighbor's birdbath. The water source can be simple, like a shallow dish or a chicken waterer.

ROAD BLOCK

When bees forage from a standing water source, they could potentially drown. Shallow water dishes should be filled with pebbles. You could also float small sticks on the surface of the water to give them something to land on.

Raising Bees Without Land

Your city's zoning laws may or may not allow this, but bees are perfectly happy with their hives located on balconies or rooftops. Be sure your roof can support you and a hive—see "Rooftop Gardening" in Chapter 4 for more information on safely using rooftops.

If you don't have access to your rooftop, check with local businesses to see if they're interested in hosting a beehive—it's a "green" thing to do, which can be good for their business. There's even a possibility that you could put your bees on the top of city hall—it's happened in Chicago and Vancouver!

Another option is to find someone who's willing to host your hive on his or her land. Bees are wonderful pollinators, so it shouldn't be difficult to find a gardener or a community garden that would love to have you keep your hive on its property. Bees need less frequent care than any other kind of food-producing animal, so it's less of a hardship to have them located away from your home.

Installing Your Bees

The specifics of opening your hive for the bees vary depending on whether you're using a Langstroth, top-bar, or other hive design. If you get your bees from a breeder, be sure to read his or her instructions before installing.

The basic procedure begins with tapping the bottom of the bee box against a hard surface. This causes the bees to settle on the floor. The queen will be in a separate small cage, which is plugged with a candy. Hang the queen cage inside the hive with the candy side up. Then you can pour your bees from the box into the hive. The bees will be docile and will go where their queen is. They will gradually eat away at the candy that's plugging the queen's cage, freeing her.

If you have a feral swarm that was captured, all the bees are clustered around the queen. Dump the swarm into the hive. As long as the queen gets in there, all the other bees will follow.

SMALL STEPS

The best time of day to install your bees is in the late afternoon or early evening, when they're more docile.

Feeding Your Bees

Bees feed themselves with the nectar and pollen they collect. However, sometimes beekeepers supplement with a sugar syrup, dry sugar, or honey. Feeding practices vary widely among beekeepers. Some never feed and say they've never lost a hive, while others believe in feeding regularly—and others fall somewhere in between. The advantage of using honey as a starter feed (instead of sugar) is that sugar can change the pH of the hive and affect the health of the bees.

When you first install your bees in their hive, there's no honey or nectar in there to sustain them. So it's a good idea to give them food until they've built up honey stores. In the fall, after harvesting, you'll leave some honey behind to get the bees through the winter. However, if for some reason there isn't enough to sustain the bees, that would also be a good time to feed them. If the weather is too cold for syrup, you could use dry sugar instead.

Feeding isn't without its risks, which is why many beekeepers do it only when absolutely necessary. The food attracts pests and can trigger "robbing," when outside bees try to enter the hive to get food. A variety of bee feeders are available, depending on the type of hive you're using.

Hive Maintenance

As mentioned earlier, the worker bees do most of the tasks involved in taking care of the hive, so there's not much for the beekeeper to do. Your job basically boils down to one of inspecting to be sure things within the hive are functioning the way they should.

Veteran beekeepers typically only open their hives six to eight times from early spring to late fall. (The hives typically don't need to be open during the winter.) In general, try to disturb your bees as little as possible. However, new beekeepers will probably want to inspect their hives more frequently, sometimes as often as once per week, to learn about the bees and familiarize themselves with the workings of the hive. It's best to open the hive on sunny days between 10 A.M. and 5 P.M., when most of the bees are out foraging.

SMALL STEPS

Bees are extraordinarily sensitive to smells and don't take kindly to strong odors. Your bees will be less disturbed by your presence if you wear clean clothes, have recently showered, and don't wear perfume or cologne while inspecting your hive.

The queen is the one bee the hive can't do without, so your main goal should be to make sure your queen is present and healthy. If you can spot her, that's terrific, but she's not always easy to find. Sometimes breeders mark the queen with a bit of paint, which makes her easier to identify. The simplest way to know your queen is there is to look for eggs in the cells. They're quite small and are shaped kind of like a grain of rice, but they're not difficult to spot once you learn what you're looking for.

After your hive has been settled in for a little while, you'll want to look to see that they're building their comb and filling some of the cells with liquid (which could be nectar or water) and pollen. You'll also check for any signs of disease or other problems (see the "Challenges" section later in this chapter.).

Your bees will take steps to winterize their hive by bringing in propolis to fill any cracks. You'll need to be sure your bees have enough honey to make it through the winter. The amount they need varies depending on where you live, and talking with other beekeepers in your area is the best way to get guidance. In general, if you live in colder areas, leave around 60 pounds of honey in the hive; in milder climates, your bees can get by with 30 to 40 pounds.

You should provide a windbreak for your hive if you expect a harsh winter. Some bee-keepers put a mouse guard over the entrance to the hive before winter to prevent mice from nesting in the hive.

Your queen will only be productive for a couple of years. Some beekeepers choose to purchase a new queen every fall and introduce her into the hive. However, you can let the hive take care of this naturally with a process called *supersedure*. When the queen gets older and her pheromone production decreases, the worker bees will raise a new queen to take her place. They do this by feeding one of the developing worker larvae extra *royal jelly*, which causes her to become a queen.

DEFINITION

Supersedure is the process by which a bee colony replaces an old or inferior queen with a new queen. **Royal jelly** is a nutritious substance secreted by worker bees and fed to all larvae in a colony. Feeding extra royal jelly causes a larva to develop into a queen bee.

Challenges

Having a beehive or two in your backyard will likely be fun and rewarding, but keeping bees is not without its challenges. Let's look at some solutions for problems you may encounter.

Swarming

Swarming is a natural instinct in your bees, driven by their desire to reproduce. Swarming occurs when the hive develops a new queen and then splits in half—one queen flies away with some of the workers to form a new colony. The presence of multiple enlarged, peanut-shaped cells (called swarm cells) can be a clue that your hive is thinking of swarming.

The disadvantage of swarming is that it decreases your hive's productivity because you have fewer workers to gather nectar. On the other hand, swarming is a way for honeybees to increase their numbers, which is helpful to the overall environment, especially considering the prevalence of Colony Collapse Disorder (more on this coming up).

Some beekeepers think swarming is a response to hives that become too crowded, so adding supers (when using a Langstroth hive) or empty bars (when using a top-bar hive) helps prevent it. Other beekeepers believe swarming will happen regardless of hive size. One idea for reducing swarming is to wait and harvest honey in the early spring rather than in the fall so the bees spend the spring—when they're typically more likely to swarm—focused on building more comb and storing up honey.

Colony Collapse Disorder

Colony Collapse Disorder (CCD) is a phenomenon in which the adult worker bees, or sometimes an entire colony, in a hive abruptly disappear. The term was coined in 2006 in response to a rising number of disappearing bee colonies in North America.

Honeybees are important for crop pollination, and therefore they're intimately tied to our food supply. The cause (or causes) of CCD are not fully understood. Simply by maintaining a healthy hive, you're helping combat the effects of CCD by bolstering your area's bee population.

 ROAD BLOCK

Regardless of what may be the cause of CCD, you should avoid using any pesticides on your property—and try to convince your neighbors to do the same. The increased use of pesticides worldwide is a problem for honeybees.

Pests

Most colonies will have a certain amount of mites. However, if they reach the level of an infestation, they will cause problems for the bees. Two kinds of mites are known to infect beehives. The first is the Varroa mite. Symptoms of a Varroa mite infestation are brown or red spots on larvae, deformed bees, or the sudden death of a colony in late fall. You can treat Varroa mites with a chemical miticide or use a mite-control technique that involves dusting the mites with powdered sugar.

ROAD BLOCK

If you give your bees medication of any kind while there's honey in the hive, that honey will become inedible for humans. If you're using a Langstroth hive, remove the supers before giving the bees medication.

Tracheal mites are difficult to diagnose without dissecting the bee. If you have a colony suddenly die during late winter or early spring, you might want to arrange to have your state apiary (beehive) inspector examine your bees for tracheal mites.

Ants can sometimes be a problem for bees. A healthy colony can deal with a few ants here or there, but if the ants overrun the hive, the bees may abandon it. You can stop ants from entering your hive by sprinkling lots of cinnamon around it.

Sometimes the worst pests to bother beehives are other bees. They may come to the hive in an attempt to rob it of honey or sugar syrup. You can decrease the likelihood of robbing by not leaving honey out in the open and by handling sugar syrup carefully. Even the smallest spill can attract other bees.

Diseases

A handful of diseases can affect honeybees—none of which is transferable to humans. In an effort to combat these diseases, some beekeepers medicate their colonies in the spring and fall. Others prefer not to routinely medicate, believing it's of limited value and preferring instead to focus on preventing diseases through proper hive management. The best way to decide which approach fits you best is to spend time talking with veteran beekeepers about their experiences.

Stings

Honeybees are generally gentle creatures and typically won't sting unless threatened or startled. However, it's not reasonable to think you can go your whole life as a beekeeper without getting a few stings. Most people experience mild pain and swelling when stung. Taking an antihistamine and applying a baking soda paste to the irritated area can help.

If you're allergic, a sting could be fatal. If you've never been stung before or believe that you might be allergic, it's a good idea to visit your doctor and get tested before acquiring bees. Even if you're not allergic, it's wise to keep an emergency sting kit (like an EpiPen) on hand in case you have a visitor who gets stung.

URBAN INFO

Don't forget that bees range up to 3 miles in each direction from the hive—that's more than 60,000 acres! You can share that information with any neighbors who are concerned about an overabundance of bees on their property. The bees will venture far and wide—much farther than your neighbor's backyard.

Harvesting from Your Hive

Bees are wonderfully prolific creatures. Although you may not get a lot of honey from your hive during its first season, an established hive can produce 60 to 80 pounds of honey per year!

The traditional time to harvest honey is in the fall, although some prefer to do it in the spring. There are a number of ways to harvest honey from your hive, and a whole bunch of tools are available to help you with it. While some beekeepers invest in equipment for their honey harvest—like an extractor, which uses centrifugal force to spin the honey out of the combs—you can choose a simpler technique.

One option is to place a wire cake rack in a large baking dish. Set the cut honeycomb on the rack, and let the honey drain into the dish. Another way is to use the "crush and strain" method, in which the honeycomb is mashed and then strained through a filter (like a fine screen mesh) to extract the honey. Some beekeepers even prefer to just pack the entire honeycomb into a container and store it that way.

However you choose to harvest the honey, remember to leave some behind when you harvest so your bees can eat through the winter!

Honey is the main "crop" you hope to get from bees, but they also produce beeswax, propolis, and pollen. The cappings bees place on top of the comb cells are made of beeswax. You can trim the caps off the comb before honey harvesting to save the wax. The combs are made of wax, too, but beekeepers often try to preserve the comb and return it to the hive after honey harvesting. Beeswax can be used for making candles and natural beauty products.

SMALL STEPS

Langstroth hives typically produce more honey than top-bar hives. This is partly because the comb on Langstroth frames can be spun in an extractor and then put back in the hive for the bees to reuse. It's simpler to harvest honey by draining it without an extractor, but the comb can't be reused.

If you'd like to collect propolis from your bees, you can purchase a "propolis trap" to put in your hive. Propolis is extremely sticky in warm temperatures, but if you put a full propolis trap in the freezer, it will become brittle and break off in chunks. Or, if you're up for it, you can just scrape the sticky stuff from the inside of your hive when you open it. Propolis has antimicrobial qualities and can be used in tinctures and ointments.

Traps to collect pollen from your bees are also available. Pollen is high in protein and is used as a nutritional supplement. If you have allergies, eating small amounts of local pollen each day can help reduce your symptoms. (Be sure to consult your doctor before trying this.) The honey you collect from your hive already has a little pollen in it, so you'll get some of the benefits even if you don't harvest the pollen separately.

The Least You Need to Know

- Bees do most of the work involved in maintaining a hive.
- You'll need to invest in supplies to get started, but you can choose a hive design to fit your needs.
- You should monitor your bees periodically to be sure the hive is functioning properly.
- Harvesting honey doesn't have to be complicated, and you can produce enough for your household with just one hive.

Aquaponics: Raising Fish and Plants Together

In This Chapter

- How aquaponics works
- The best system for urban homesteaders
- Getting everything set up
- Keeping your fish and plants healthy

Urban homesteading often focuses on doing a lot with just a little space, especially when it comes to food production. Therefore, it's not surprising that aquaponics is becoming more and more popular because it allows homesteaders to grow two types of food—plant crops and fish—in one system.

Aquaponics is the integration of aquaculture (raising fish in a controlled environment) with hydroponics (growing plants in a soil-less media). Because aquaponics systems are completely contained aboveground, they have numerous applications in city settings. Not only can apartment dwellers without land raise both plants and animals using aquaponics, it's also a way to produce food on concrete or land that's contaminated.

In this chapter, I explain the basic principles of aquaponics and offer advice on how you can create a system that works for your space. I talk about the maintenance needs of an aquaponics setup, the challenges you may encounter, and the harvesting of your fish and crops.

How It Works

Aquaponics is both a marvel of technology and a re-creation of a simple natural system. If you were to raise fish alone, you'd have to deal with the buildup of their waste, which can quickly become toxic. If you were to raise plants alone in a hydroponic system, you'd need to find a reliable source of fertilizer or your plants wouldn't survive. Aquaponics blends these two needs beautifully by raising both fish and plants together in a closed system.

The water from the fish tank is circulated up to the vegetable grow bed(s). The naturally occurring bacteria in the system convert the bad stuff in the fish waste (ammonium and nitrites) into good stuff that the plants can use for fertilizer (nitrates). The plants absorb these nutrients and filter the water, which is then returned to the fish tank.

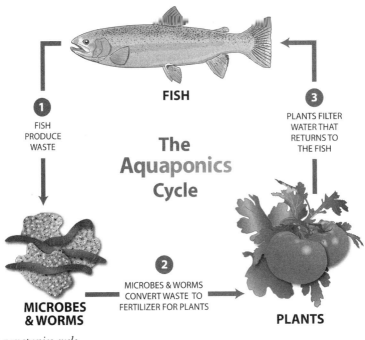

The aquaponics cycle.
(Courtesy of The Aquaponic Source)

It sounds simple, and aquaponics gardeners are quick to say it's easy because the system does most of the work. As often happens with the best conceived ideas in food production, aquaponics is mimicking what happens in nature. The symbiotic relationship between fish and plants can be found in lakes and streams all over the world. Creating a system like this in your apartment or backyard takes a little effort. However, aquaponics growers—whether they raise food commercially or just for their family's consumption—feel the time spent setting up and maintaining their system is well worth it.

Aquaponics gives homesteaders the opportunity to grow protein and vegetables together in one integrated system. Although setting up the tank requires an initial "investment" of water, growing vegetables with aquaponics actually uses only 10 percent of the water you'd use to grow the same crop in the ground. There's no need to purchase fertilizer for your aquaponics garden because the fish provide it. And because the plants aren't growing in soil, there are no weeds or soil-borne diseases to worry about. Plus, it's a way for city dwellers to raise animal protein without worrying about waste disposal or noise. It's getting more and more difficult for consumers to purchase fish grown or harvested sustainably, so having the ability to produce your own is helpful.

SMALL STEPS

Aquaponics systems take some work to get set up, but once they're running, part of their beauty is that they're largely self-maintaining. You should plan on spending a few minutes two or three times a day to feed your fish and monitor their health, plus an additional few minutes each week for water testing.

What to Raise?

The potential harvest from an aquaponics system is surprisingly diverse. A number of different fish and plant varieties can successfully grow together using aquaponics.

Types of Fish

The following varieties of fish do well in an aquaponics tank:

- Tilapia
- Perch
- Catfish
- Peruvian pacu
- Oscars
- Trout
- Koi (decorative, not for eating)
- Freshwater prawns

Generally speaking, it's best to limit yourself to one type of fish per tank because each species of fish may prefer a different water temperature and pH level. However, if you acquire a couple of different kinds of fish that thrive in the same conditions—and won't eat each other—they can share a tank. By far, the fish most commonly chosen for aquaponics systems is tilapia. Tilapia are fairly hardy—they're reasonably tolerant of fluctuations in water temperature, pH, and oxygen levels. They're easy to breed and eat an omnivorous diet. Tilapia grow to harvest size, about 1 to 1.5 pounds, in about nine months. They produce white-fleshed meat.

ROAD BLOCK

Tilapia are such good breeders that they're considered invasive pests in many areas. It's not uncommon for the State Fish and Game Department to make it illegal to stock a private pond with tilapia. However, they're generally allowed in closed-loop systems like aquaponics.

Purchasing Your Fish

The most cost-effective way to stock your aquaponics system is to purchase young fish, called fingerlings. If you have a fish farm in your area that can sell you fingerlings to get you started, that's great. If not, you'll need to look online to find a hatchery that will ship. Many places are used to catering to large commercial farmers, but a few will sell a small number of fingerlings to individuals. (See Appendix B for a hatchery that is able to ship small quantities.)

Tilapia fingerlings cost $1 to $2 each, less if you're ordering in large quantities. How many fish you order depends on the size of your tank—I talk about this more in "The Right Design for Your Space" section later in this chapter.

You'll also need to consider whether you'd like to order only male tilapia, which grow faster, or both sexes so your fish can breed and replenish themselves. See the "Breeding" section later in this chapter for more information.

Growing Crops

An aquaponics system provides a nutrient-rich environment for vegetables and fruits, allowing gardeners to plant using the intensive spacing outlined in Appendix C. Food crops commonly grown in aquaponics include the following:

- Salad greens—lettuce, watercress, spinach, etc.
- Microgreens—radish and turnip sprouts, baby greens
- Herbs—especially basil
- Chives

- Broccoli

- Tomatoes

- Peppers

- Cucumbers

- Squash—both summer and winter

- Melons

- Strawberries

These are the more popular aquaponics crops, but they're far from the only possibilities. When you're first starting with aquaponics, it may be wise to plant items known to be successful, but once you become comfortable with your system, feel free to experiment!

Leafy green plants have lower nutritional requirements and can be grown in a newly developing aquaponics system. Plants that yield fruit (like tomatoes and squash) have higher nutritional needs and require a well-stocked fish tank to support their growth.

The Right Design for Your Space

Aquaponics growers have a number of design options to choose from. There's the raft system, in which the plants are grown in floating styrofoam boards in their own tank, separate from the fish. There's also a method known as Nutrient Film Technique (NFT). This involves designing long, narrow channels through which the water flows and in which the crops are planted. Both of these techniques are utilized in commercial aquaponics because they can produce high yields of specific vegetables. However, they also usually require a significant amount of money and upkeep.

For the purpose of urban homesteading, let's focus on the simpler media-filled bed method of aquaponics. This system is relatively inexpensive to build, is easy to maintain, and allows you to grow a wide variety of vegetables.

The Basics of the Media-Filled Bed Method

A media-filled bed is pretty much what it sounds like—a vegetable bed filled with growing media. The media is typically ¾ inch gravel, expanded shale, or something called expanded clay pebbles. Regardless of what you choose as your growing media, the secret ingredient in media-filled beds is red wriggler worms. These are the same little critters used in worm composting (see Chapter 22). They work to break down the solids that make their way into the grow beds, thereby keeping things nice and clean. Plus, red worms excrete their own fertilizer, which gives an additional boost to the plants!

Here's an example of a media-filled bed setup.
(Courtesy of The Aquaponic Source)

Crops are either transplanted as seedlings or seeded directly into the growing media. The ideal depth for a grow bed in this system is 12 inches. Grow beds can be made out of pretty much any container, provided it's waterproof and made of food-grade or food-safe materials. Rubbermaid stock feeding tubs (available at farm and stock supply stores) are a popular choice. If you're using plastic, look for polypropylene (labeled PP) or polyethylene (labeled PE or HDPE).

You have a variety of options when looking for something to hold your fish as well. If you have a smaller system, a large household aquarium will work. You can also use a stock feeding tank or even an old bathtub!

URBAN INFO

Although most aquaponics gardeners fill their grow beds with nonsoil growing media, one method of aquaponics incorporates compost into the grow beds. Developed by Growing Power, this system also has shallower grow beds and is best for growing microgreens and watercress.

You'll need some kind of plastic tubing to connect your fish tank to the grow bed, and a water pump to move the water. Many aquaponics growers prefer to use the "flood and drain" approach to cycling their water. By imitating natural systems like waves or tidal surges, it allows the plants' roots to receive needed oxygen. To create these conditions, you'll need to attach a timer to your water pump. The typical approach is to have the timer run the pump for 15 minutes and then shut off for 45 minutes. The goal is to cycle all the water in your tank each hour.

Homesteaders who want to try aquaponics have a number of choices if they'd like to build their own system. You can purchase a comprehensive kit, which contains everything you need plus instructions for putting it together. If you'd like a less-expensive option, you can find plans online for creating your own system from scratch. You can purchase items to serve as the fish tank and grow bed, or you can use reclaimed items (provided they're made of materials that are safe for fish and humans). Some aquaponics growers like to use salvaged 55-gallon plastic barrels to create both their grow beds and their fish tanks. Known as "barrel-ponics," this design can provide home growers with a productive system that fits well in a city backyard.

Size Calculations (and Raising Fish Without Land)

First and foremost, you want a system large enough to grow edible fish. Aquaponics systems can be small enough to fit on a desk, but those are meant to hold goldfish (which, of course, aren't very good eating). Beyond that, the size of your setup is dictated by the space you have available. Consider whether you want your system to be kept inside or outside. If you put it outside you can have a larger setup, but you'll need to take steps to protect your fish and plants during the winter months (see the "Winter Care" section later in this chapter).

There are a couple of size guidelines you can use when planning your system. The first is the density of fish that's feasible per gallon of water. Opinions on this vary across the aquaponics community. Commercial tilapia growers stock at a density as high as one fish per gallon of water, but many home growers feel that a ratio of one fish per 5 gallons of water is more appropriate for a simple media-filled bed system. Keep in mind that when the fish are young, they're significantly smaller, so more can fit into a given space. However, as the fish grow, crowding—and a buildup of fish waste—becomes more of an issue.

Veteran aquaponics growers say the most common mistake people make is overstocking their fish tank. When you're first starting your system, it's wise to begin with less fish density so you can be sure your plants adequately process the fish waste. Once your system has been running for a while, you can increase the density—as long as you're monitoring the water quality and removing fish when they reach harvest size.

The second guideline is the relationship between the size of the fish tank and the size of the grow bed(s). Remember that the plants do the critical work of filtering the water so it can be safely returned to the fish tank. If there aren't enough plants to absorb what the fish excrete, the water will become contaminated and the fish may not survive.

A ratio of 1:1 between your tank and your grow bed is a good way to ensure you have enough plant life to adequately clean your water. This ratio can be difficult to figure out because tanks are typically measured in gallons and grow beds are typically measured in cubic feet. Thankfully, there's a formula you can use to determine how large your grow bed should be, based on the size of your tank:

Tank size = 50 gallons

The ratio between the tank and the grow bed should be 1:1.

Therefore, for a 50-gallon tank, the grow bed should be 50 gallons.

7.48 gallons = 1 cubic foot

50 gallons ÷ 7.48 = 7 cubic feet (rounding up)

Therefore, a grow bed that's 3.5 feet long × 2 feet wide × 1 foot deep (14 cubic feet) is sufficient to support a 50-gallon fish tank.

Getting Started

Once you get your system set up, it's tempting to want to dump in your fish, plant your vegetables, and get everything going immediately. However, that approach generally doesn't yield good results. There are some steps you need to take to get your system ready to go.

Preparing the Water

You may be surprised to learn that one of the most important living organisms in your aquaponics system isn't fish or plants—it's bacteria. The bacteria that reside in your grow bed media convert fish waste (ammonia and nitrites) into nutrients that plants can use (nitrates).

These nitrifying bacteria don't magically appear as soon as you plug in your water pump, but they will be attracted to your system if you provide them with a source of ammonia to feed on. There are a few ways to accomplish this, but the approach that generally works the best, and results in the fewest fish casualties, is called "fishless cycling." This means you add a little ammonia to your system daily, and then let it run for a couple of weeks without any fish.

You can use a variety of ammonia sources, but the safest option is to use pure ammonia. It is also called "clear ammonia," and is available in the cleaning section of the grocery store.

SMALL STEPS

Some aquaponics growers don't feel the need to go far to locate a source of ammonia to jump-start the nitrification process. They just pee into their fish tank water!

In addition to giving the good bacteria time to accumulate, fishless cycling provides another benefit. It's okay to use municipal water from your outdoor spigot or your tap to fill your tank. However, fish and bacteria don't like chlorinated water, and there's chlorine in your city's drinking water. You could add chemicals to your tank to remove the chlorine, but that's not necessary considering chlorine disperses on its own over time if it's exposed to air. So the time your system spends in fishless cycling also allows the chlorine to dissipate.

There's one additional thing critical to the health of your fish: oxygen. Be sure to provide an air pump in the tank to oxygenate the water for the fish; otherwise, they can die after just 30 minutes without it. Turning on the air pump before your fish arrive ensures that the water is sufficiently oxygenated.

After a couple of weeks, you can test your water to see if it's ready for the fish. The water testing kit most commonly used by home growers is the Freshwater Master Test Kit sold by Aquarium Pharmaceutical Incorporated (available at local aquarium stores or online). You'll be testing for pH and the presence of ammonia and nitrites. Your ideal pH is 6.8 to 7.0, but anything in the range of 6.4 to 7.6 is fine. You can use products called pH Up or pH Down to adjust your pH if necessary. Your ammonia and nitrites need to be almost gone (below 0.5 parts per million each) before it's safe to add the fish. If they're too high, it means the fishless cycling needs to go on a little longer. When all the numbers add up, you're ready to move ahead!

Planting Your Garden

If you'd like, you can plant your crops during the fishless cycling phase. This gives them an opportunity to get established before they're responsible for filtering water for the fish. However, without fish in the tank, there's no fertilizer in the water, so your plants won't grow nearly as quickly. If you want to give your plants a little boost, you can use a soluble seaweed called Maxicrop during the fishless cycling phase, or you could add an extra handful of worm compost to the growing media.

When you're first starting with an aquaponics system, it's a good idea to focus on leafy greens until the bacteria become truly efficient at converting fish waste into plant food. Follow the guidelines presented in Chapter 5 for direct seeding and/or transplanting your crops. You can use the spacing indicated in Appendix C.

Introducing Your Fish

If your fish are being delivered through the mail, time your order so they don't arrive until after your fishless cycling is complete and your system is ready to go. Be sure your water temperature is in the proper range for the fish you're using. (For tilapia, it's 72°F to 78°F.) You might have to get a heater for your tank to bring the water temperature into the correct range.

It's best to float your bag of newly arrived fish in your fish tank for about 15 minutes to allow the fish to gradually adjust to the water temperature. It's also a good idea to add some of your tank water to their traveling bag, in case there's a difference in the waters' pH.

Caring for Your System

Much of the work required for aquaponics occurs on the front end, with setting up your system and preparing the water for your fish. When everything is up and running, it's surprisingly easy to maintain.

Feeding the Fish

Fish need to be fed two or three times a day. Those who raise fish commercially have come up with a formula for how much they need to eat—1 to 2 percent of their body weight per day—but it doesn't have to be that complicated. Feed your fish only what they can eat in 5 to 10 minutes. Uneaten food will cause water quality issues, so be sure not to overfeed.

URBAN INFO

Fish are stellar converters of feed into meat. It takes 1.7 pounds of fish feed to produce 1 pound of meat. Compare this to the amount of feed chickens (2.4 pounds), turkey (5.2 pounds), and beef (9.0) need to consume to produce 1 pound of meat.

Tilapia should be given feed that's 25 to 35 percent protein. Many commercially available feeds, including Fish Chow, meet that requirement. Tilapia are omnivores, so their diet can be supplemented with certain plants. A good choice is duckweed, a fast-growing aquatic plant. Duckweed floats on the surface of water and can be grown in the fish tank. However, it grows rapidly, so you'll need to monitor it so it doesn't take over the water space!

Maintenance

The most important maintenance task you will do, next to feeding the fish, is testing the water periodically. The frequency of testing is up to you, but it's a good idea to test more frequently at first and then less as time goes on. You might begin by testing every other day and then decrease it to once every week or so when your system is well established.

You should test for pH, ammonia, nitrites, and nitrates. The ideal levels for the first three items are described in the "Preparing the Water" section earlier in this chapter, and you don't want to let your nitrates climb above 150 parts per million.

ROAD BLOCK

If you find you need to adjust your tank's pH, be sure to make changes very slowly. Fish are sensitive, and they can be stressed by rapid changes in pH. Don't adjust any faster than 0.2 points per day.

It's also important to monitor the water temperature of your tank. The best way to do this is to keep a waterproof thermometer in the water at all times, so you can easily read the temperature when you need to.

The beds in a media-filled system should be able to stay clean on their own—provided the beds are 12 inches deep and contain some red wriggler worms to eat up the waste solids. However, you still may want to clean your plumbing from time to time. Don't use any cleaning agents—not even mild dish soap—on your equipment because fish are so sensitive. Some people even choose to use nonchlorinated water when rinsing out their tubes because any chlorine present will kill the beneficial bacteria inside the tubes. However, if you need to use a little tap water, things will likely be just fine; there's enough bacteria in the rest of the system to keep things going. Speaking of which, the walls of your fish tank may become slimy. However, this slime is composed of that good bacteria, so wiping the tank walls is not recommended.

You can tend the crops you raise in your grow bed in the same way as plants in any container garden. See Chapter 4 for tips on maintenance and troubleshooting. The one element of container gardening you won't have to worry about is fertilizing—the fish take care of that for you!

Breeding

Many commercial tilapia farmers raise only male fish because they grow to harvest size significantly faster than females. However, this approach requires that the grower pur-chase new tilapia fingerlings a couple of times each year. While it's certainly possible for a homesteader to follow this method, you can try a simpler, and less costly, approach by allowing your fish to breed and increase their numbers on their own. Tilapia reach sexual maturity at around 11 weeks old, and from that point on, they'll *spawn* every 5 or 6 weeks.

Tilapia are great breeders—so much so you'll quickly find yourself with far more baby fish than you need. This will create a serious problem for your system if it's not managed because the overabundance of fish will produce more waste than your plants can filter. Also, overcrowding will prevent your maturing tilapia from reaching their full harvest size. The solution is to regularly *cull* fish from both ends of the size spectrum—harvest the fish that have reached a good size for eating, and remove most of the just-hatched fish from the tank. These very small baby fish (called fry) can be given to someone who is starting his or her own aquaponics system. They also make great chicken food, or they can be added to compost (see Chapter 22 for more information).

DEFINITION

Spawning is the production of large quantities of eggs in water—the means by which fish and amphibians reproduce. **Culling** is the process of removing animals from a group based on specific criteria. Culled animals are typically killed and usually eaten (unless they're removed from the group due to illness).

Common Challenges

If you follow the steps laid out in this chapter when you're setting up your system, prepar-ing the water, and managing breeding, you shouldn't experience a lot of problems. Fish and plants thrive when they're protected from stress. Just be sure your water temperature stays constant, the fish are fed regularly, and the water quality is what it should be (which you'll know if you test it periodically).

The quickest way to throw a wrench in a well-functioning system is a power failure, which would affect both the water pump and the air pump. It's a good idea to invest in a backup power source for your system because a brief period without electricity could lead to extensive fish death.

You should monitor your fish to be sure they're staying healthy. Signs of problems include the following:

- Gasping at the surface or gulping for air

- Changed swim pattern

- Decreased or stopped eating

- Change in coloring

- Blemishes or parasites

Gasping at the surface is an indication that the fish don't have enough oxygen in their water, so the air pump is not working sufficiently. Changes in swim pattern, eating habits, or coloring are likely an indication of stress, so you should be sure the water temperature and quality are satisfactory. If a fish is demonstrating signs of illness (like blemishes or parasites), it should be taken out of the tank and destroyed.

Other than a curious housecat stalking an indoor aquarium, you're not likely to encounter problems with urban predators. Most outdoor fish tanks are sufficiently deep to prevent predators from reaching in a paw and snagging a fish. You'll notice that fish are aware of anyone (or anything) that approaches their tank, and they typically "hide" in the corners or along the bottom until the intruder has passed.

Winter Care

Both fish and plants need special care to protect them from freezing temperatures. If you have an outdoor system that can be moved inside during cold months, that's the easiest way to ensure year-round growing. Systems left outside need to be sheltered by a structure, like a greenhouse. You can maintain the correct water temperature for the fish with a water heater, but the heat given off by the water may not be enough to protect the plants. In that case, you may need to heat the greenhouse.

ROAD BLOCK

Be sure the surface of your fish tank never freezes over. Exposed water is essential for the oxygen exchange fish need to stay healthy.

Tilapia need fairly warm water to thrive. In that regard, trying to raise tilapia in the winter is analogous to raising tomatoes in January—it can be done, but it's not the natural crop for the season. Some aquaponics gardeners prefer to switch to fish that can tolerate cooler water, like perch, during the winter months. However, this requires completely removing all the current fish and restocking the tank with the new variety so it will halt a breeding program.

Harvesting

Fish harvesting is a type of animal processing that can reasonably be done in a city kitchen. Tilapia can be butchered in the same way fisherman do with other types of fish. Some people think the way to ensure the best flavor is to place the fish in a separate tank, without food, for two to four days prior to harvesting. If you don't have the space (or resources) for a separate tank, you can just harvest the fish straight out of the main tank. Fish meat freezes well and can also be dried in a dehydrator.

The crops in your aquaponics system can be harvested following the guidelines in Chapter 9. It's best to stagger your crop harvests so there are always functioning plants in the system to clean the water.

The Least You Need to Know

- Aquaponics is the growing of fish and plants together in a closed system.
- A wide variety of plants, and several types of fish, can be raised successfully using aquaponics.
- Aquaponics systems vary in size and can be small enough to fit easily inside an apartment.
- A well-balanced system should be easy to care for because the fish and plants work together to keep things clean and healthy.

A Homemade Life in the City

Part

4

Much of what constitutes homesteading was on its way to becoming a lost art. You may have fond memories of your grandmother canning apple butter or making goat milk soap, but not many of us had the chance to learn at her apron strings. Luckily, there's been a revival in these homesteading skills, and it's being led by city dwellers.

As more people begin to garden, they want to learn how to preserve the food they grow so it lasts past the end of summer. Raising backyard food-producing animals means more milk to be made into cheese (and sometimes more meat for soup). Plus, urban homesteaders are discovering that creating their own beauty products and cleaning supplies is not only cheap and fun but safer and better for the environment.

In this part, we delve into the things that make up the homesteading arts so you can create a homemade life for yourself in the city!

Small-Batch Food Preserving

Chapter 16

In This Chapter

* Canning safely and easily
* Creating fermented vegetables
* Different ways to dry foods
* Creating an urban root cellar

There's nothing like fresh vegetables, fruits, and herbs in the summertime, but what about after the frost has come and your garden's no longer producing? You can continue to feed yourself from your garden (or the farmers' market) long after all the plants have died away. There's something wonderfully hopeful about putting up food for the future, and a little work done up front can make for easier meal preparations down the road. Plus, opening a jar of home-canned spaghetti sauce in January is a great way to bring a bit of summer to a cold winter day.

Don't be fooled into thinking you need a big, country-style kitchen to preserve food. Eugenia Bone is one of the foremost experts on modern-day preserving, and she does it all in her New York City apartment!

There is a wealth of ways to preserve your harvest. In this chapter, I talk about four methods particularly well suited to homesteaders with limited space: canning, fermenting, drying, and root cellaring. I also provide information on the basic procedures, foods that can be preserved, and what you need to get started. This helps you decide which method (or methods) you'd like to use, and then you can use the books listed in Appendix B to find specific recipes for preserving.

Canning (Even in a Tiny Kitchen)

Canning may seem intimidating at first, but it's actually quite simple and straightforward once you get the hang of it. Some would-be home canners also worry about becoming sick from spoiled food. But as long as you follow a few basic safety precautions, you can proceed with confidence (see the "Keeping It Safe" section later in this chapter).

If you're only going to can one thing, consider canning tomatoes. Tomatoes are fun to grow, easy to can, and useful in lots of recipes. Plus, commercially canned tomatoes contain BPA (Bisphenol A), which many people avoid because it has been linked to a variety of health problems.

Canning in small batches isn't necessarily a disadvantage—it allows you to focus more on variety than quantity. Every canning recipe can be safely scaled down if you'd prefer to produce just 4 pints of something instead of 12. If you clean as you go, you can do all the work of canning—including chopping the food, filling the jars, and setting them out for cooling—in a 2×2-foot counter space.

You could get by with just one stove burner if you needed to and still can whatever you'd like. All the equipment you'll use for canning is relatively small, except perhaps the pot. Even those are available in city-kitchen-friendly sizes—you can find small canning pots meant to process 4 pints at a time. Luckily, the only requirement for storing canned goods is that the space be relatively dark and cool. So if you fill up your kitchen cupboards, you can find space for your extra canned food under your desk, tucked in a bookshelf, or even in your closet!

SMALL STEPS

Your canning pot can serve other purposes when it's not being used on the stove. It's a great (and rodent-proof!) place to store your bags of flour or grains.

Basic Equipment

The type of pot you'll use for canning depends on whether you're doing water bath or pressure canning. I discuss the different pots in the later "The Two Types of Canning" section. Regardless of which method you use for canning, a few tools make it faster, safer, and easier:

Canning jars: Sometimes called Mason jars, these glass jars are made to withstand the heating and cooling pressures of the home canning process. Glass jars not designed specifically for canning won't work.

Rings and lids: The tops of canning jars come in two pieces. The rings are metal bands that screw onto the top of the jar, holding the lid in place. The lids are flat metal discs with a rubberized rim that forms a seal during canning. Lids are good for one use only. You can purchase lids separately because you can reuse your jars and rings.

Canning funnel: This funnel with a handle and a wide opening fits perfectly inside regular mouth canning jars. It's very helpful for filling the jars without making a mess.

Jar lifter: These special tongs are used to transfer jars in and out of the canning pot. The lifter is essential for safely moving the heavy jars out of very hot water—regular kitchen tongs won't do the trick.

Spatula: A narrow, nonmetallic spatula helps release air bubbles that are trapped in the filled jars before sealing.

Magnetic wand: This isn't essential, but it's helpful for lifting sterilized lids out of hot water. It's basically a stick with a magnet attached to the end. You can purchase these or make one yourself with a paint stirrer and a couple of loose magnets.

You can find all these items through Craigslist, Freecycle, and garage sales. Canning jars are an especially popular garage sale item. However, you'll likely have the most luck getting your hands on used canning equipment in the spring. In the fall, during "canning season," there will be a lot of competition, and the equipment will be harder to find. If you'd like to purchase these items new, the best place to buy them locally is probably the hardware store. The "big box" hardware stores don't carry these items, but the smaller stores do, especially in the fall. You can also find canning equipment online and in some grocery stores.

The Scoop on Ingredients

The most important thing to remember when canning can be summed up in a few words: "garbage in = garbage out." Don't think you can use old, wrinkly, or otherwise undesirable produce in your canning and end up with good results. Generally speaking, you want to select the freshest produce and use it at its peak. It's true that farmers' markets sometimes sell boxes of "seconds" (usually tomatoes, peaches, or apples) to be used in canning. These should be items that have just suffered cosmetic damage, not food that's old or rotting.

Canning recipes usually call for some of the following ingredients:

Salt: Regular (iodized) table salt should not be used in canning because it can cause cloudiness and discoloration. The best choice is pickling salt, which you can find in most grocery stores and sometimes in hardware stores, next to the canning supplies. You can use kosher salt in a pinch, but because it's coarser than canning salt, you'll need to use a little more than what's called for in the recipe.

Lemon juice: Always use bottled lemon juice when canning. The acid level of fresh lemon juice is variable and can't be relied on, but the bottled juice is consistent.

Vinegar: Distilled white vinegar is the best choice, unless something different is specifically called for in a recipe.

ROAD BLOCK

Vinegar corrodes many types of metals, so only use glass, stainless-steel, ceramic, or enamel containers and utensils when canning anything that contains vinegar.

Sweeteners: Sugar is sometimes used in small amounts for flavor, like in condiments and pickles, but in jams and jellies, it works as both a gelling agent and a preservative. Raw or cane sugar can be used instead of processed sugar. Low- or no-sugar jam recipes are available, but they rely more heavily on pectin or gelatin. Some recipes tell you that you can substitute honey. If so, use only $\frac{1}{3}$ to $\frac{1}{2}$ as much honey because it's sweeter than sugar. You can also use stevia or other sugar substitutes, if you prefer.

Pectin: Available commercially at grocery or hardware stores in either powdered or liquid form, pectin is used to thicken jams, jellies, and spreads.

The Two Types of Canning

The first type of canning, and the one that's easiest for beginners, is called water bath canning. This method can only be used for processing "high-acid" foods, or foods with a pH of 4.6 or lower. Fruits like apples, grapes, peaches, cherries, and plums; tomatoes; and anything pickled are good candidates.

SMALL STEPS

At first glance, it may seem that your food options are pretty limited if you want to do water bath canning. However, within the realm of high-acid foods, there are lots of choices and variations. Spaghetti sauce, spiced peach jam, pickled zucchini, chutney, and apple butter can all be processed in a water bath.

With water bath canning, you place filled jars in a pot of boiling water for a specified period of time. The heat destroys the bacteria and molds that would spoil the food, and the lids form an airtight seal when the jars are removed from the boiling water. Water bath canning can be done in any pot, as long as it provides enough space to cover the jars with an inch of water, plus an extra couple of inches so the water has room to boil.

You need to have some kind of a rack to keep the jars from touching the bottom of the pot. Canning pots come with racks (so do tamale-steaming pots), but you can also purchase a round cake cooling rack and use it for this purpose.

Low-acid foods (including all vegetables except tomatoes, plus meat, poultry, and seafood) must be processed in a pressure canner. Acid guards against bacterial growth, so without it, these foods need the added heat pressure canning provides.

A pressure canner is *not* the same thing as a pressure cooker, so don't try to substitute one for the other. A variety of pressure canners are available, but they all have a way of regulating the pounds of pressure produced within the canner. Read the instructions thoroughly, and follow the manufacturer's directions when using a pressure canner.

How to Do It

Regardless of whether you can in a water bath or a pressure canner, many of the steps are the same:

1. Sterilize the canning jars. Fill the canner with water, place the empty jars in the pot, and bring it to a boil for a few minutes. (After the jars are removed and filled with ingredients, the same water can be used for the processing.)

2. Sterilize the tools that will be touching the food, including the lids, funnel, and ladle. You can do this in a separate pot of boiling water (if you have room on your stove) or in the main canning pot.

3. Prepare the ingredients according to the recipe directions. Wash all produce thoroughly, and cut away any blemishes.

4. Place the ingredients in the jars (which still may be hot from the sterilization), but don't fill them all the way. The amount of space between the rim of the jar and the top of the food is called the headspace, and it should be specified in your recipe—it's usually around $^1/_2$ inch. This space is important because it gives the food room to expand while it's being processed without interfering with the jar's seal.

5. Wipe the rim of the jar with a clean paper towel. Place the lid on the jar and screw on the ring until it's gently secure but not completely tight. This is some-times called "fingertip tight."

6. Process the jars in the water bath or pressure canner according to the recipe's instructions.

7. After processing is complete, remove the lid and allow the jars to cool briefly in the canner. Wait 5 minutes for water bath or 10 minutes for pressure canning.

8. Lift the jars without tilting them, and set them on the counter. The water pooled on the top of each lid will evaporate as the jar seals.

9. You'll know the jars have sealed when the little button on the lid pulls down. You'll hear it happen. Each one makes a little "plug" as the jars cool. Ti mun times takes several hours for jars to seal, so be patient.

Jars that have not sealed can be fitted with a new lid and reprocessed, but it's usually best to just put them in the fridge and eat the contents before they spoil.

Keeping It Safe

As long as you take care to keep things clean and follow the canning recipe precisely, you shouldn't need to worry about unsafe food. However, it is important to understand that canning is not like other kinds of cooking—it's not okay to improvise with canning. The recipes have been formulated to ensure that the processed food will be safe, and if you tinker with the ingredients, it could have serious consequences. Bacterial contamination—including botulism, which can be fatal—isn't always easy to see or smell. Therefore, it pays to be careful up front. If you discover that any of your jars have broken seals, are leaking, contain mold, or smell strange, discard the food.

ROAD BLOCK

Exercise caution when using canning recipes from the Internet. Anyone can post a recipe, but that doesn't mean it's been tested for the proper pH and is safe to use. It's best to get your recipes from the USDA or a well-respected canning book (see Appendix B).

Canning recipes list times and pressures formulated for sea level. If you live at altitude, you always need to make adjustments. For water bath canning, increase the processing times as indicated in the following table. Pressure canning doesn't require an increase in processing time, but you may need to increase the pressure. When pressure canning using a weighted gauge, add the 15-pound weight if you live 1,000 feet or more above sea level. If you're pressure canning using a dial gauge, increase the pressure by 1 pound for each 2,000 feet above sea level.

Altitude Adjustments for Water Bath Canning

Feet Above Sea Level	Increase in Processing Time
1,000 to 3,000	+5 minutes
3,001 to 6,000	+10 minutes
6,001 to 8,000	+15 minutes
8,001 to 10,000	+20 minutes

Fermenting

Fermenting with salt is one of the oldest methods of preserving vegetables. Also known as lacto-fermentation, the process uses lactic microbial organisms to create an environment so acidic that food-spoiling bacteria can't survive.

Fermented vegetables can keep for a short while without refrigeration, especially if they're stored in a cool place. You can extend the life of the vegetables significantly by putting them in the refrigerator, and you can also process them in a water bath canner.

Cabbage is probably the most popular kind of fermented vegetable, but you can use this method with most any firm vegetable, including turnips, radishes, carrots, zucchini, eggplant, cucumbers, garlic, and baby onions. You can also ferment green beans, although they need to be blanched (quickly steamed) first.

 URBAN INFO

You're likely already familiar with another product of lactic acid fermentation—yogurt! Fermenting enthusiasts state that many of the beneficial bacteria and enzymes present in yogurt are also available in fermented vegetables.

Two Approaches to Fermenting

Although the basic premise of fermenting is salt and water, there are a couple of ways to go about it. The first is by creating a *brine*. The exact proportions vary a bit depending on the recipe, but it's often a ratio of around 3 tablespoons salt to 1 quart water. The vegetables are submerged in the brine, and the process of fermentation begins in about three days.

DEFINITION

A **brine** is a mixture of salt and water used to preserve food. It may also contain spices, sugar, and/or vinegar.

The other method is to mix a vegetable directly with salt, a process known as "dry salting." The salt pulls the water out of the vegetable, which creates a concentrated brine. This is the process used when fermenting cabbage-based foods, like sauerkraut and kimchi.

How It Works

Fermentation can take place in anything from a quart jar to a 10-gallon crock. The container can be made of stoneware, glass, ceramic, stainless steel, or food-grade plastic. It's important for the container to be as sterile as possible before beginning fermentation. If you're using a Mason jar, you can submerge it briefly in boiling water, as you do when canning. Nonboilable containers can be run through a hot dishwasher or cleaned with a white vinegar solution.

Follow the recipe directions for preparing the ingredients, and fill the container. Leave a few inches of headspace below the rim because the process of fermenting creates gas bubbles, and you don't want the brine overflowing onto your counter. It's important that the vegetables stay completely submerged in the brine to prevent spoiling. To accomplish this, weigh down the vegetables. The simplest way is to fill a sealable plastic bag with brine (not water—use brine in case the bag springs a leak) and place it on top of the submerged vegetables. Cover the open container with cheesecloth or a dish towel. This keeps out flies but still allows the air circulation needed for fermentation.

For the first couple of days, the brine will remain clear. Around the third day, it will start to become cloudy, and you'll notice gas bubbles forming. Scum will begin to form on top of the brine. It's made of yeasts and mold that feed on the lactic acid, so you'll need to remove it daily.

The total processing time for fermented foods varies—anywhere from 10 days for standard pickles to 4 to 6 weeks for sauerkraut. You can sample the food periodically, and when you decide it's "done," it can be done. A good cue that the food is ready is when the gas bubbles cease forming. However, fermentation is a living process, and it will continue as long as you keep the food at room temperature. Refrigeration slows things down; canning stops it completely.

ROAD BLOCK

Be aware of the signs of spoilage. If the fermented food becomes soft, slippery, slimy, or discolored, don't eat it.

Aside from the vegetables themselves, which should always be fresh and unblemished, salt and water are the most important ingredients in fermenting. Always use noniodized pickling salt. (Follow the same salt guidelines described for canning. See "The Scoop on Ingredients" section earlier in this chapter.) If your tap water is good to drink, it should be good for fermenting. However, hard or overchlorinated water doesn't work well. If your water stains the toilet, it's too hard, and if it smells like bleach, it's too chlorinated. If this is the case, boil your tap water for 15 minutes and let it stand for 24 hours. Remove any scum that forms on the top of the water. Carefully ladle the water out of the pot, leaving behind any sediment in the bottom.

Drying

Drying is a simple way to preserve food, and it can be used effectively for a wide range of items. Fruits, vegetables, herbs, and even poultry and meat can be effectively preserved using drying. There are a number of different techniques to choose from when drying, although not every drying technique works for every type of food.

A few general principles hold true, regardless of what you're drying or which method you're using. The first is that good circulation is key. Drying is usually done on baskets, trays made of wooden slats, or stainless steel screens. Solid trays like cookie sheets typically don't work well because they block the airflow.

ROAD BLOCK

Don't make your drying trays out of any material that will contaminate the food. Refrain from using aluminum, copper, vinyl, fiberglass, non-food-grade plastic, or galvanized metal.

Also remember that it's important to get things dry. It's the lack of moisture that prevents the growth of bacteria. Fruits and meats should be leathery; vegetables and herbs should be dry enough to break if squeezed; beans should be hard but not wrinkled. The pieces of food may need to be rotated periodically to ensure even drying.

Last but not least, be sure you can see the process through to the end once you begin. Partially dried food can quickly become contaminated with bacteria.

Prepping and Storing Dried Food

As with any kind of preserving, you should select the freshest and best food for drying. Choose items that are free of blemishes, and be sure to clean everything thoroughly. The thinner you slice something, the faster the drying process will be. Light-colored fruit like apples and peaches can be sprinkled with ascorbic acid or powdered vitamin C to prevent discoloration. Vegetables can be peeled if you'd like. Meat or poultry that's been partially frozen is easier to slice and handle.

Store dried food in an airtight container in a cool, dark place. If you notice any condensation in the container, the food needs additional drying. If you dried your food using one of the outdoor methods (see the next section), it's likely that an insect laid eggs on the food. There's nothing like hatched larvae to render food unappetizing, so you may want to put the dried food in the freezer for a week or two before storing it. This step is called pasteurization, and it will kill anything that may have gotten into your food.

Methods of Drying

Each method of drying food has its advantages and challenges. Some can be done very simply, and others require an investment of time or money.

Air drying: This works best with herbs, onions, garlic, and "dry" beans (as opposed to green beans). It can be done either inside or outside, and basically it involves hanging the food out of direct sunlight until it's dried. With onions and garlic, you can braid their tops and hang them from the braid. It's best to tie herbs inside a small brown paper bag when drying. The bag will protect the herbs from dust and catch any leaves that fall from the stem. Beans are even easier—you can allow them to dry on the plant and then shell and store them.

Car drying: This is a favorite method of urban homesteaders because it harnesses something that's typically a problem—cars heating up in the summer sun—and turns it into an asset. This method works best with higher-acid foods like fruit and tomatoes and quick-drying foods like kale and collards. Park the car in the sun and close the windows. (There will still be adequate airflow in the car, and you don't want critters getting to the food.) Spread the food in a single layer on the tray, and cover it with cheesecloth or muslin. Place the food in the car, out of direct sunlight.

ROAD BLOCK

You'll need to walk the line between too hot and too cool when drying food. The ideal temperature is around 140°F. If you allow things to get too cool (below 100°F), bacteria will begin to grow, but if it gets too hot (above 180°F), the food will cook instead of dry.

Solar drying: A number of plans are available online for building a solar dryer. Some are a simple "hot box" design, similar to a cold frame (see Chapter 9) but with screened holes in the sides and bottom to facilitate airflow. Another option is the popular Appalachian model, which effectively harnesses the sun's power to allow you to dry several trays of food at once. See the resources in Appendix B to find online plans for building a solar dryer. They work well for the same foods as car drying, but if yours is well constructed, you can also use it for low-acid vegetables.

Oven drying: This method has the advantage of working even when the weather is wet or cloudy, but it does use energy and can only dry a couple of trays of food at a time. You can use the oven to dry fruits, vegetables, and meat. You'll need to use a wooden spoon to keep the oven door propped open slightly to provide good airflow. The temperature in the oven should be between 140°F and 160°F.

Electric dehydrators: Available commercially, these allow you to dry several trays of food at once. You can easily control both the heat and airflow of these dehydrators, and they can be used to dry any kind of fruit (including making fruit leathers), vegetables, or meat.

Root "Cellaring"

We rejoice when the vegetables and fruits we grow become ripe. However, the same process that causes produce to ripen is what causes it to rot. Root cellaring is the method of halting, or at least slowing down, the natural decomposing of ripe food.

For the purpose of this book, I'm using "cellaring" in quotations because most urban homesteaders don't have an honest-to-goodness root cellar—or even a basement for that matter! Even so, there are several things you can do to utilize the principles of root cellaring to preserve your harvest. One of the best aspects of this method of food preservation is that it doesn't require energy in the form of heating (like firing up the canner on the stove) or cooling (like refrigerating fermented vegetables).

Before getting into the specifics of root cellaring, it's important to note that the best way to store food is often just to leave it in the garden. Root vegetables (except potatoes), cooking greens, and salad greens can take some degree of frost and cool weather. A true hard freeze would ruin root vegetables, but it's not uncommon for gardeners to dig through the snow to harvest their kale. The exception to this rule is often winter squash. It's not that winter squash can't handle a little cold; rather, it's a tempting treat for squirrels and other critters, so it's usually best to bring in winter squash sooner rather than later. (See Chapter 9 for more information on when to harvest garden vegetables.)

The Basics

The most important aspect of root cellaring is keeping temperatures cool. The ideal range is between 32°F and 40°F. You need to maintain a balance between keeping the temperatures cool enough to preserve the produce but not so cold that things freeze, which will cause them to rot. You'll need to take steps to keep your stored food protected from rodents, too.

Also, as with every other method of food preservation, you should only store produce that's in good condition. Eat the damaged food now and save the good stuff for storage.

SMALL STEPS

The old adage that "one rotten apple spoils the whole barrel" is very true, and it's equally valid for potatoes.

True to name, all root vegetables—including potatoes, beets, turnips, parsnips, and carrots—are good candidates for root cellaring, as are garlic and onions. Apples and pears can also be stored. Even green tomatoes will last for a while in storage, although they'll ripen along the way.

Generally speaking, produce will keep longer if it's not allowed to touch. Whenever possible, store "dry" items—potatoes, onions, garlic, apples, pears, and tomatoes—in layers of straw shredded newspaper. The exception to this is winter squash, which can do just fine piled in a corner. "Moist" items—beets, turnips, parsnips, carrots, and radishes—like to be layered in damp, not soggy, sawdust or sand. Unless you live in a very humid climate, you cannot successfully store "moist" produce in the open air. They'll quickly become withered.

Inside Options

If you want to store food inside, begin by finding the darkest, coolest part of your home. If you have a spot that's notoriously drafty, use it to your advantage! "Dry"-loving items can be stored (in their newspaper layers) in ventilated boxes, mesh or burlap sacks, or even laundry baskets. You can pack "moist" items into a plastic tub or a cooler with their damp sawdust or sand. It's best to leave the lid open to prevent mold.

SMALL STEPS

Try hanging your bag of potatoes or apples from the ceiling—it will keep them out of reach of mice.

Some people build their own root cellar by putting up a couple of insulated walls around a window or vent, which can then be left open during the winter to lower the temperature of the little room. This is an ambitious project, but it may be worthwhile if you have construction skills and want to make a long-term investment in food preservation. See the resources in Appendix B to find plans for building a root cellar.

Outside Options

If you want to store food outside, get ready to start digging. The protection of the ground, in addition to the insulation you'll provide, helps regulate the temperature and keeps things from becoming too hot or too cold. You'll need to have a secure, critter-proof container such as a metal trashcan or a cooler for holding your food.

Dig a hole deep enough to sink your container into the ground. Layer the food with insulating material—either dry straw or moist sand, depending on what you're storing—and be sure the top few inches of the container are just insulation. Attach the lid securely. You don't need to bury your container completely with dirt, but covering it with a little straw helps keep it cool. Opening the container as infrequently as possible also helps maintain a good temperature.

ROAD BLOCK

This may seem like common sense, but don't dig a big hole for root cellaring unless you own the land, or have written permission from your landlord. A root cellaring version of guerrilla gardening (using someone else's land without permission—see Chapter 6) isn't likely to work out well.

The Least You Need to Know

- It doesn't take a lot of space to successfully can your harvest.
- Fermenting takes time but very little effort and can be a great way to preserve your vegetables.
- You can dry food using the sun's energy, among other easy methods.
- You can create a root cellar (or something like it) for storing vegetables, even in an apartment!

Preparing What You Harvest

In This Chapter

- The basics of soft and hard cheeses
- Homemade yogurt and butter
- Using meat and vegetable scraps for stock
- Herbed vinegars and oils

The bounty of an urban homestead doesn't need to be limited to vegetables and fruits. Many homesteaders are also raising food-producing animals in the city and facing the wonderful challenge of making good use of what the animals provide.

In this chapter, I talk about cheese-making as a way to use and preserve milk, including a few simple cheese recipes you can prepare with ingredients you probably already have in your kitchen. I also cover what you need to know to make homemade yogurt and butter, plus I share recipes for meat- and vegetable-based stocks. In addition, I explore how you can use your garden's herbs to further enhance your meals.

Cheeses

Cheese—milk's leap toward immortality.

—Clifton Fadiman

Many home cooks will happily bake a loaf of bread but would never dream of trying to make cheese. For some reason, cheese-making has a reputation for being mysterious and difficult, when it's actually quite simple in most cases.

Cheese has been a way to preserve and transport milk for thousands of years—ever since humans started keeping domesticated animals. At its most basic level, cheese-making is the separation of milk into its *curds* and *whey*. You can add special bacteria to the cheese culture to produce different flavors and can press and age the curds to create a hard cheese, but those steps are all variations on a theme. Technically, once you have curds, you've created cheese.

 DEFINITION

Curds are the solids formed during coagulation of milk. They contain the milk's fat and most of its protein. **Whey** is the watery part of milk left behind when the curds are removed. It contains the milk's water, sugar, minerals, and some protein.

You can begin your adventures into cheese-making with the soft cheeses, which are easier to make, require fewer ingredients and less equipment, and have the advantage of being edible right away. When you have a few good soft cheeses under your belt, you can try a hard cheese—but you'll have to wait around three months for it to age before you can give it a taste!

There are a number of reasons why homesteaders might want to have cheese-making as a part of their repertoire, but it's an especially important skill for anyone raising dairy animals. There's only so much milk you can drink, so using the milk to make cheese is a great way to utilize your animals' production.

About the Milk

You don't need to have a couple of goats in your backyard to make your own cheese (although it helps). Cheese made from home-raised milk is fantastic, but you can also make great cheese using store-bought milk.

When preparing to make cheese, there are a few things about milk to keep in mind:

Milk fat: In almost every case, whole milk is the best choice for making cheese. You could technically use 1 percent or 2 percent milk, but your yield (the amount of cheese you get for the amount of milk) will be less, and the quality will be lower. A couple of cheeses—like Parmesan and some recipes for cottage—turn out better if they're made with skim milk. (If you're milking your own animals, the milk you get from them is whole milk.)

 URBAN INFO

One gallon of whole milk yields about 1 pound of cheese.

Ultra-high pasteurized: Unless you live in one of the few states that allow it, you cannot purchase raw milk from the store. Pretty much all milk sold commercially has been pasteurized, which is fine for cheese-making. However, milk that's been ultra-high pasteurized (sometimes marked as UHP or "ultra pasteurized") doesn't work for almost all cheeses. You can make simple cheeses—like paneer—but most other cheeses won't turn out correctly with UHP milk. Instead, look for milk labeled "pasteurized." Unfortunately, sometimes UHP milk is mislabeled as pasteurized, so it can take a little trial and error to find the right brand. If you follow the cheese-making directions precisely and don't get good results, try changing brands of milk.

Homogenized: Milk purchased from the store is homogenized, which means the fat globules have been broken up so they'll be evenly dispersed. There are no problems with using homogenized milk to make cheese. If you're using fresh (backyard) goat milk, it's naturally homogenized.

Goat and cow milk: Most cheese recipes that call for cow milk can also be made with goat milk. Some cheeses (like chèvre) specifically require goat's milk. Remember that goat milk becomes more "goaty" the longer it sits. If you're purchasing goat milk from the store, it has been out of the goat for quite some time and will likely produce a fairly strong-tasting cheese. Cheese made from fresh goat milk is much milder (unless you take steps to intentionally strengthen the flavor).

Frozen: You can use frozen milk to make cheese, but the curd will be more fragile, and the flavor won't be as good as cheese made with fresh milk.

Dry milk powder: You can make cheese from this. 1⅓ cups dry milk powder dissolved in 3¾ cups water makes 1 quart milk.

If you have access to home-raised milk, either from your own backyard animals or another local source, you're all set for your cheese-making. The milk you use must be from a healthy animal, it must be completely clean, and it must have been kept refrigerated. If you wouldn't feel safe drinking the milk, you shouldn't use it to make cheese. If you want, you can pasteurize the milk before making your cheese (see Chapter 12 for instructions).

Part of what makes different cheeses unique is the type of food the animal grazes on. Keep this in mind when feeding your animals—don't give your goats onions or garlic, unless you plan on making a savory feta!

Soft Cheeses

You can make a wide variety of simple soft cheeses at home, from mozzarella to chèvre to cream cheese. Some take as little time as 20 to 30 minutes to make! Many soft cheeses, although easy to make, require the addition of certain cultures or bacteria. You can purchase these from a local cheese-making shop or online (see the website in Appendix B).

If you're looking for more instant gratification, there are a handful of cheeses you can make using ingredients you probably already have on hand. I've given you recipes for three cheeses in the following sections.

SMALL STEPS

Once you're comfortable making simple soft cheeses, you can try your hand at a number of mold-ripened cheeses like Brie and Camembert.

Chenna

Chenna has a creamy texture. It's great in savory vegetable dishes and can also be used in desserts.

Here's how to make chenna:

> ½ gal. whole milk
>
> ¼ cup fresh lemon juice (if using bottled, increase amount slightly)

In a heavy-bottomed pan over medium heat, slowly heat the milk, stirring frequently. Bring the milk just to a foaming boil, reduce heat to low, and immediately (before the foam subsides) drizzle in the lemon juice while stirring. Continue stirring gently until the curds begin to separate from the whey. If curds don't form after a minute of stirring, add a little more lemon juice.

Remove the pot from the heat, cover, and set aside for 10 minutes to allow the curds to solidify. Strain through a wire strainer or a colander lined with cheesecloth, pressing firmly with your hands to remove as much of the whey as possible.

SMALL STEPS

When you make cheese, always save the whey. You can use whey as a soup base, in smoothies, to ferment vegetables, and even as an animal feed—chickens love it!

Paneer

Paneer is a cheese that commonly appears in Indian cooking, but it can be used in all kinds of cuisine. You can crumble paneer and add it directly to a dish, or you can cube and pan-fry it.

To make paneer, use the same ingredients and preparation directions as for chenna. After draining out the whey, leave the cheese in the strainer and set a few small plates on top of the cheese to act as a weight. Place the cheese in the refrigerator for 45 minutes to press.

SMALL STEPS

If you don't have cheesecloth handy, you can use an old (very clean) T-shirt to strain your cheese.

Simple Cottage Cheese

To make cottage cheese, follow the directions for paneer. After draining out the whey and before putting it in the refrigerator to press, rinse the curds briefly with cool running water. Press the curds for 45 minutes, as with the paneer. Crumble the pressed curds, salt to taste, and store it in the refrigerator.

Just before you eat your cheese, add a little cream to the salted curds. Don't add the cream ahead of time. It's better if you wait until right before eating.

Hard Cheeses

Creating hard cheese is part science and part art. You'll need a number of supplies to make hard cheeses, including bacterial starters, *rennet*, cheese wax, and a cheese press. I offer resources for purchasing these supplies, along with a great book to teach you how to make hard cheese, in Appendix B. Most hard cheeses take three or four months to age, and you'll need a place to store them where you can control the temperature. Successfully creating your own hard cheeses can be challenging, but it's a feat well worth attempting.

DEFINITION

Rennet is an enzyme that coagulates milk. It can be obtained from animal or plant sources.

Purchasing a cheese press can be a significant investment. Luckily, it's something a resourceful homesteader can make fairly easily. Basic materials include nontreated wood boards, dowels, washers, screws, weights, a pie pan (optional), and a cheese mold or PVC pipe. You'll also need an electric drill with wood boring bits and a screwdriver. Appendix B lists a website that has plans for building a cheese press.

A homemade cheese press.
(Courtesy of Fias Co Farm)

More Fun with Dairy

Cheese isn't the only homemade dairy product you can create in your kitchen. Using the milk from your backyard animals (or the store), you can make yogurt and butter to add to your meals.

Yogurt

Yogurt is milk that's been cultured thanks to the work of friendly bacteria, similar to the fermentation of vegetables (see Chapter 16). It can be made with either goat or cow milk, and the milk can be whole, reduced fat, or skim. Some recipes recommend adding dry milk powder to get a thicker yogurt, but it isn't necessary. To get your yogurt going, you'll need some starter bacteria, which you can easily obtain by purchasing a small amount of plain yogurt from the store or ordering a starter culture online.

The best part of making yogurt is that, once you've created your own, you can save a small amount to use as the starter for the next batch. Unfortunately, this doesn't work indefinitely, because at some point unwanted bacteria will get into your starter and produce off-tasting yogurt. When that happens, just begin again with a little fresh yogurt from the store.

ROAD BLOCK

If you're making yogurt with goat milk, you might think you should purchase goat milk yogurt to use as your starter culture. However, store-bought goat products are often fairly sour tasting, so you'll get a better end product if you use cow milk yogurt as your starter.

Here's how to make yogurt:

1 qt. milk

2 TB. yogurt or starter culture

In a saucepan over medium heat, bring the milk to 180°F, stirring frequently. You can use a cooking thermometer to check the temperature, but if you don't have one, 180°F is about when the milk starts to foam and swell in the pan—but before it boils. (If you're using a yogurt maker, it may hold more than 1 quart of milk, so adjust the recipe accordingly.)

Remove the milk from the stove and cool. You can speed up the cooling process by setting the saucepan in a dish of ice water. Once the milk has reached 110°F, stir in the yogurt or starter culture, and mix thoroughly.

Place the mixture in a sterilized jar (or jars). Keep it at around 110°F for 7 hours (10 hours if you're using skim milk). After the yogurt has formed, refrigerate it for up to 2 weeks. A quart of milk will give you about a quart of yogurt. You can flavor the yogurt with jams, fresh fruit, maple syrup, honey, or spices like cinnamon and nutmeg. Don't forget to save a little plain yogurt to make the next batch.

SMALL STEPS

It is helpful to use a cooking thermometer to check the milk's temperature before adding the yogurt. Milk that's too hot will kill the beneficial bacteria, and milk that's too cool may not set properly.

Making yogurt is remarkably simple. The trickiest part is keeping the yogurt at 110°F for 7 hours so the bacteria can do their work and thicken the milk into yogurt. A small electric yogurt maker can make this task very easy. These cost around $35 and maintain the perfect temperature for yogurt. However, if you don't want to spend money on a new gadget—or try to find a place to store it!—the following are some ideas for keeping your yogurt mixture warm. (Be sure to keep the mixture in a heat-proof container like a canning jar for all these ideas.)

Heating pad: Place a heating pad inside a large pot. Turn the pad to medium, cover the pad with dishtowels, and place the jar of yogurt in the pot. Put additional towels on top of the yogurt jar for insulation and put the lid on the pot.

Cooler: Put a few inches of hot water into a small insulated cooler. Put the yogurt jar into the cooler, and secure the cooler's lid. Check the water temperature periodically, adding more hot water as needed.

Slow cooker: You'll use the same setup with the slow cooker as with the cooler. However, you'll need to cover the jar and open the slow cooker with a pillow because the slow cooker lid won't fit over the quart jar.

Butter

Butter is made by agitating cream. That's pretty much all there is to it! In fact, many home cooks have accidentally created butter by overwhipping cream when making whipped cream. Plus, as you make your own butter, you'll get homemade buttermilk as a by-product of the butter-creation process. You can use buttermilk in biscuits or pancakes, and some people also enjoy drinking it.

The first step in butter-making is to get the cream. If you don't have access to fresh (nonhomogenized) milk, you can purchase cream from the store. If you're using fresh cow milk, the cream will separate from the milk after about 24 hours of sitting in the refrigerator. Carefully spoon off the layer of cream that forms on the surface, leaving the milk behind.

If you're using fresh goat milk, getting the cream is a little trickier. Goat milk is naturally homogenized, so it's more reluctant to separate. You can purchase a cream separator that spins the cream out of the milk for you, but they're quite expensive.

A little natural separation happens in goat milk, and you can take advantage of it. Pour your milk into a large, shallow pan, and leave it uncovered (and undisturbed) in your refrigerator for 24 hours. Then carefully scoop out the cream. Using this method, you'll need about 5 gallons goat milk to accumulate 1 pint cream. There's nothing wrong with doing it a little bit at a time and storing the cream in the freezer or the fridge until you have enough to make butter.

Once you have a pint of cream, you're ready to begin. Some people like to leave their cream out at room temperature to "ripen" for 8 hours or so before making butter, and others choose to use it straight from the fridge. You can try it both ways and see which flavor you prefer.

URBAN INFO

If your goat milk cream has been accumulating in the refrigerator for several days, it will already be a little ripe.

You can agitate the cream using an electric hand mixer, a stand mixer, or a food processor. You can even do it the low-tech way, shaking it in a glass canning jar. Be sure to use a jar that's twice as large as the amount of cream so it has plenty of room to agitate—use a quart jar for a pint of cream. Mix, blend, or shake the cream until butter solids form and separate from the buttermilk. If you're doing this with a mixer, your cream will initially turn into whipped cream, but if you keep going, it will become butter.

Drain and reserve the buttermilk. Knead the butter gently with a spatula or wooden spoon, forcing out the buttermilk. Once it seems that there's no more milk in the butter, add cool water. Agitate gently, kneading the butter to release any remaining buttermilk. Drain the water and repeat until the water runs clear. If any milk is left in the butter, it will spoil. Salting butter helps it last longer, so if you'd like, you can fold in a little salt to taste at this point. Store the butter, covered, in the refrigerator.

A pint of cream yields about ½ pound butter and 1 cup buttermilk.

DIY Stock

Stocks are an example of frugality and recycling at their finest: using the uneaten parts of an animal (the bones) to create something that's both delicious and nutritious.

Good stock is a staple in every gourmet restaurant kitchen. You can use it to make wonderful soups, sauces, and reductions. Plus, properly prepared stock contains numerous minerals and other healthy things—there's a reason why chicken broth is recommended as a remedy for the flu! Vegetables can be added to meat stocks to enrich the flavor or used on their own for vegetarian stock. Well-made stock is the very definition of a "slow food" … the longer you let it simmer, the richer it will be.

Chicken Stock

You can make chicken stock with either a complete chicken or just chicken parts like the bones and skin. If you use the whole chicken, you'll get a two-for-one: great stock plus cooked chicken meat for sandwiches or salads. However, you can still get terrific stock by using whatever's left over after you've had your chicken dinner. Or if you'd prefer, you can purchase some of the less popular chicken parts (like necks and backs) from your butcher for a discount and use those for your stock.

Here's how to make chicken stock:

> 1 whole chicken, a couple pounds chicken parts, or leftover bones of a roast chicken
>
> 1 gal. cold water
>
> 2 TB. vinegar (distilled white vinegar is best)
>
> 1 large onion, chopped
>
> 2 carrots, chopped
>
> 3 celery sticks (including leaves), chopped
>
> 3 cloves garlic, unpeeled and unchopped (optional)
>
> Few springs fresh thyme (optional)
>
> 1 bunch fresh parsley

If using a whole chicken, cut it into several pieces. Place the chicken and cold water in a large, stainless-steel pot. Add the vinegar, onion, carrots, celery, garlic, and thyme. Let stand, covered, for 30 minutes to allow the vinegar to draw some of the nutrients out of the bones.

Place over high heat and bring to a boil. Reduce heat to low and simmer, covered, for 12 to 24 hours. Periodically skim off any froth that gathers on the broth's surface. Add the parsley about 10 minutes before the stock is finished cooking.

Strain the stock through a fine mesh colander or a regular colander lined with cheese-cloth. Refrigerate the stock immediately. The fat will rise to the surface as the stock cools. You can skim this off and used it as a sauté fat for cooking.

Rabbit or Vegetable Stock

You can make rabbit stock by following the chicken stock recipe. A whole rabbit typically weighs a little less than a whole chicken, but that's not a big concern. Ultimately, it's the length of time you cook the stock, not the amount of meat or bones it contains, that has the biggest impact on its flavor.

A basic vegetable stock uses the vegetables and herbs listed in the chicken stock recipe—onions, carrots, celery, garlic, parsley, and thyme. To give this stock a richer flavor, sauté the chopped vegetables in 1 tablespoon olive oil until they're lightly brown. Add the water, and simmer. (You can leave out the vinegar when making vegetable stock.) Cook for a couple hours before straining and refrigerating.

Don't feel limited to only the vegetables listed in the chicken stock recipe. Feel free to add bits and pieces—including the peels and trimmings—from other vegetables you might have lying around.

Avoid most of the vegetables in the cole family, like broccoli, cauliflower, and cabbage (turnips are okay), because they'll make the stock stinky. Also, tomatoes will change the character of the stock and make it less versatile, so it's better to save the tomatoes for tomato soup.

Storing Stock

Freshly made stock can keep for five days in the refrigerator or for several months in the freezer. It's a good idea to freeze your stock in serving-size portions like a pint or a quart so you can thaw only what you need for a recipe.

To make the best use of your freezer space, remove the stockpot lid for the last several hours the stock is simmering. The stock will cook down, and the flavors will become concentrated. You can then freeze it in even smaller portions and add water after it's thawed to get your desired quantity of stock.

SMALL STEPS

Be sure to label your containers of stock because they'll all look the same when frozen. Also include the date on the label so you can use the oldest stock first.

Flavoring with Herbs

Any urban garden, no matter how small, has room for herbs. They require very little attention, thrive in the ground or in containers, and often return from year to year. Herbs can be a reliable part of your harvest, so it's beneficial to learn ways to utilize herbs in your cooking and preserving.

Infusing Vinegars and Oils

Chopping up a handful of herbs and throwing them into a dish adds flavor to one meal, but preserving those same herbs in vinegar or oil extends their usefulness for weeks. There are few other things you can do that are so simple yet add so much pizzazz to your food. Herb-infused vinegars can be used on salads or steamed vegetables. Herb-infused oils are great for sautéing meats or vegetables or drizzling over finished dishes.

SMALL STEPS

Herb-infused vinegars and oils make wonderful homemade holiday gifts.

You should always infuse vinegar in a glass jar with a secure lid. Be sure your container is clean, dry, and sterilized. The best option is to use a canning jar that's been immersed in boiling water for several minutes.

When it comes to your herbs, you can use either dried or fresh. Citrus peel or garlic is also a good choice. Fresh herbs should be clean, with any bad parts removed.

Distilled white vinegar competes less with the flavor of the herbs, but many people prefer the taste of apple cider vinegar. Red or white wine vinegar is also an option.

What herb and vinegar combination you use is up to you, but here are some suggestions to get you started:

- Sage and tarragon in distilled white vinegar

- Basil in apple cider vinegar

- Garlic, mint, and lemon peel in white wine vinegar

Here's how to put it all together:

Place the herbs in the sterilized jar. You don't need to pack the jar—a small handful of herbs is sufficient for a pint-size jar. Be sure the herbs are pushed down into the jar so they'll be completely covered by the vinegar.

Fill the jar with vinegar. Attach the lid and set the jar in a cool, dark place or the refrigerator for 2 weeks.

Pour the vinegar through a fine mesh strainer to remove the herbs and return the strained vinegar to the jar. Store in the refrigerator for up to 4 months.

If you'd like to make an herb-infused oil, you can follow the same directions as for the vinegar, with a few adjustments:

Wash fresh herbs and allow them to completely air-dry before covering with oil. Any moisture left on the herbs will lead to bacterial growth and can cause spoilage. When the herbs are dry, tear or bruise them slightly to help release their flavor and add them to the jar.

Olive, sunflower, and safflower oils are all good choices for infusing. You can gently warm the oil over low heat before using it to cover the herbs. Herb-infused oils can keep for up to 2 months.

Here are some suggested combinations for herb-infused oils:

- Garlic and rosemary
- Basil and oregano
- Tarragon and mint

When shopping for your jars, get plastic lids instead of metal because metal lids can react with the oil and impart a metallic taste.

ROAD BLOCK

It's always a good idea to refrigerate your infused oil, but it's mandatory if you've used garlic in the infusion. Unrefrigerated garlic oil can become contaminated with botulism.

Salad Dressing and Marinade

A well-planned garden, with staggered plantings of greens, can provide you with salads from late spring through the fall. If you're able to grow greens in a window box, you can have salad all year round!

A homegrown salad deserves the freshness of a homemade salad dressing. The simplest to make—and yet infinitely versatile—is an herb vinaigrette. Whipping up a vinaigrette isn't about following a recipe as much as it is about a ratio (oil to vinegar), plus a few extra things thrown in for flavor. Once you get the hang of it, you can make your own vinaigrette using a pinch of this and a dab of that. As long as you taste as you go, the finished product will turn out fine.

Experimenting with the herbs is the best part of creating your own vinaigrettes. Feel free to dream up new combinations, or just use whatever herbs you happen to have on hand!

SMALL STEPS

The vinegar you use has an impact on the salad dressing's flavor. Balsamic and rice vinegar are sweet, apple cider and red wine vinegar are strong, and white wine vinegar is mild. Don't use distilled white vinegar.

Fresh Herb Vinaigrette

You can use any herbs or blend of herbs from your garden, including basil, oregano, sage, thyme, tarragon, savory, and/or rosemary.

Here's how to make fresh herb vinaigrette:

$^1/_2$ oz. (about a handful) fresh herbs

1 cup olive oil

2 TB. water

$^1/_4$ to $^1/_2$ cup vinegar, or to taste

$^1/_2$ tsp. salt

Pinch black pepper

Optional ingredients to enhance flavor:

$^1/_2$ tsp. sugar or honey

Pinch cayenne

Dash Bragg's Liquid Aminos or soy sauce

1 tsp. Dijon mustard

Rinse the herbs and remove any brown areas or tough stems. Add all ingredients to a blender or food processor, and mix. If you'd prefer a nonelectric option, you can mince the herbs finely and mix all the ingredients in a jar. Adjust seasonings to taste.

If you'd like a garden-fresh marinade to use for poultry, meat, fish, or vegetables, just omit the vinegar from the recipe.

The Least You Need to Know

- Soft cheese, yogurt, and butter can all be made quickly and easily in a home kitchen.
- It's possible to preserve milk by making hard cheese, but it requires an investment in time and supplies.
- Your meat and vegetable scraps can be used to make delicious stocks.
- Fresh herbs can get an extended life by combining them with vinegar or oil—or both!

The Finer Things

In This Chapter

- Step-by-step guide to soap-making
- A simple no-shampoo "shampoo"
- Easy DIY lotion
- Making your own yarn

Anyone who's done a little homesteading knows the things you grow, raise, or produce yourself are usually much better than what you can buy in the store. Homegrown tomatoes, backyard eggs, and home-canned jam all taste better than their commercial equivalents. This also holds true for things we use, too, like soaps and yarn. Making these for yourself—the items I'm calling "the finer things"—helps you sustainably produce more of what you use daily.

In this chapter, I give you simple instructions for making your own soap. I also talk about a way to clean your hair without using soap (in case your homemade soap has run out). I go over the steps for creating a simple lotion using just three ingredients and show you how you can tweak the recipe to make it your own. Plus, I go over the basics for creating yarn using materials you may have produced on your urban homestead.

Making Soap in a City Kitchen

Soap-making is one of those homesteading skills, like canning and cheese-making, that can seem intimidating from a distance but is actually quite doable once you learn the basic steps. Creating your own soap allows you to select the oils you'd prefer to use as a base, choose your favorite scents, and customize it with additions like oats or honey, depending on whether you want to use your handmade soap to clean your face, body, or hair.

You can experiment with many varieties of soap. Goat milk soap is a great option, in part because it's moisturizing and good for those with sensitive skin. Plus, it's another way for urban homesteaders to enjoy the benefits of raising backyard goats. You can store your extra goat milk in the freezer and use it whenever you're ready to make soap. Even if you're not a goat owner, you can still make this soap for yourself using powdered or liquid goat milk from the store—you can even make it with whole cow milk if you'd like.

The method that follows has been streamlined for urban homesteaders. It has fewer steps and required supplies than traditional soap making.

Lye Safety

Before going any further, it is important to emphasize the need for caution when working with lye. Lye is a caustic agent that can cause severe problems if you allow it to touch your body, particularly when it's wet. Always wear gloves and eye protection while making soap. It's also a good idea to wear long sleeves, pants, and socks. Avoid making soap when children or rowdy pets are around.

If you get lye on your body, immediately flush the area thoroughly with water, and seek medical attention if necessary.

URBAN INFO

You may have heard that vinegar is the best way to neutralize a lye burn. According to the Material Safety Data Sheet, however, adding an acid like vinegar to lye releases a great deal of heat—ouch! The recommended first-aid treatment is thorough flushing with water.

Making Goat Milk Soap

Now, on to the soap! Thanks to Amy Kalinchuk for sharing the basic steps for creating goat milk soap. For detailed photo instructions on soap-making, plus wonderful recipes, visit soapcrone.com/ebook.php.

Here's what you need to make goat milk soap:

- Newspapers and paper towels
- A plastic dish tub (12-quart size works well)
- Clear plastic food wrap
- Parchment paper or muslin cloth
- 2 old bath towels

- A big kitchen knife

- 1 large (12 cups or larger), 1 medium (6 cups or larger), and 1 small nonaluminum bowl

- 1 glass or plastic pitcher (1 quart or larger); a measuring cup with spout will also work

- An immersion blender, sometimes called a hand or stick blender (this tool is optional but very helpful)

- 2 plastic or wooden spoons

- Digital scale that can measure to the tenth of an ounce (postal scales work best)

- Disposable medical exam gloves (latex or nitrile)

- Goggles, safety glasses, or other eye protection

ROAD BLOCK

It's important to follow the guidelines regarding the types of material for your supplies. Plastic and glass are good, and stainless steel can also be used. *Never use supplies made of aluminum, because they will react with the lye.*

Now for the ingredients you'll need for goat milk soap. All measurements are by weight, so "ounce" means a weight ounce, not a fluid ounce. Most of the oils used for soap-making are sold in grocery stores, and the more exotic varieties can be found at restaurant supply stores or online. Lye (sodium hydroxide, or NaOH) is sometimes sold as a cleaning supply in grocery or hardware stores, but it can also be bought at a chemical company.

Here's what you need:

36 oz. olive oil

12 oz. coconut oil

16 oz. almond oil (can also use grapeseed oil)

8 oz. goat milk, refrigerated

8.9 oz. lye

16 oz. bottled or distilled water

2 oz. essential oils (optional)

1. Cover your work space with old newspapers.

2. Place the medium bowl on the scale. Measure out the olive oil on the scale, and dump it into the large bowl. Repeat with the coconut oil. Measure the almond oil, then set it aside in the medium bowl. (Don't add the almond oil to the large bowl.)

3. Put on your gloves and eye protection. Place the small bowl on the scale, use it to measure out the lye, and set it aside.

4. Place the pitcher on the scale and measure the water. Carefully add the lye to the water. Use a spoon to gently stir until the lye is dissolved and starts to turn clear.

   Note: The lye will release fumes when it's dissolving. Some people choose to do this step outside. You can also set the pitcher underneath your stove's hood vent while it's running.

5. Slowly pour the lye solution over the fats (olive and coconut oils) in the large bowl. Gently stir with the spoon. The lye mixture is hot and will melt the fats.

6. While the fats are melting, add the refrigerated goat milk to the almond oil in the medium bowl, and stir with the immersion blender (or a wooden spoon) until it is thoroughly incorporated.

7. When the hard fats in the tub are completely melted, add the almond oil/goat milk mixture to the lye/fat mixture in the large bowl. Blend with the immersion blender until it starts to look like thickened pudding. At this point, you can add the essential oil if you'd like and blend it into the soap. If you have a stand mixer, you can also use it (with its whisk attachment) for this step. It's possible to skip the electric appliances altogether and just mix the soap with a wooden spoon, but it will take longer to thicken.

8. Line the plastic tub with parchment paper or muslin cloth. Pour in the soap mixture.

9. Smooth the top of the soap with a clean spoon. Lay a sheet of plastic wrap on top of the soap. (The wrap should be touching the soap.) Gently lift the tub and tap it on the table to release any air.

10. Place the tub aside so the soap can set. Wrap one towel around the base of the tub, and lay another one over the top. It will take the soap 8 to 12 hours to set if you live in a dry climate or up to 24 hours if you are in a humid climate.

11. Hand wash all your equipment while still wearing your protective gloves.

12. Once the soap has set, place a layer of paper towel onto your work space. Remove the plastic wrap from the top of the soap, and turn the tub upside-down to dislodge the soap onto the paper towels. Peel off the parchment paper or muslin cloth.

13. Cut the soap into 16 bars. Place the bars back into the tub, standing up on their short ends. Allow the soap to cure until it's completely hard, which could take 3 to 6 weeks.

You may be wondering how something as caustic as lye could be safe to use in a product that's going to end up on your skin. The process of the lye reacting with the fats neutralizes it. You can make very gentle, sensitive skin–friendly soaps using this method. It is also possible to make soap without lye, using glycerin instead. However, this method consists of using a premade soap base, which is then mixed with a couple of ingredients and poured into a mold. It's much less "homemade" than a lye soap recipe.

Simple Shampoo Alternative

One of the wonderful things about homemade soap is that you can use it for washing your hair in addition to your face and body. However, there's another way to clean your hair without using any kind of soap at all. Many people use this method because homemade or natural soap is not always easy to come by, and most commercial shampoos contain ingredients that are harmful to the environment (and some people believe they can be toxic to humans, too). Plus, it's significantly less expensive to clean your hair this way rather than purchasing shampoo.

This technique is often called "no-poo," which is a shortening of "no shampoo." To do it, here's what you need:

Baking soda

Apple cider vinegar

Nonbreakable cup

The easiest way to do this is while you're in the shower, but it can also be done while washing your hair in the sink.

1. Wet your hair. Put a couple of tablespoons baking soda into the cup and add 1 or 2 cups water. Gently shake the cup or stir with your finger until the baking soda is dissolved.

2. Apply the mixture to your scalp. Gently work it in around the roots, and distribute it through the rest of your hair. Rinse with water.

3. Put a couple of tablespoons apple cider vinegar into the cup, and add 1 or 2 cups water. Pour this onto your hair, and work it in just as you did with the baking soda mixture. Rinse thoroughly.

The baking soda is a cleaning agent, and the vinegar works to condition your hair. The degree to which this method is successful depends on you and your hair type. Some people use it as the only way of washing their hair, and others just do it occasionally. Since it's likely you already have the necessary ingredients lying around, it might be worth a try if you're interested in finding an alternative to shampoo!

Making Lotion with Beeswax

It can sometimes feel like understanding—or even just reading—the ingredients on the back of a lotion bottle requires a degree in advanced chemistry. Purchasing lotion can get expensive after a while, too. Luckily, lotion-making is another one of those homesteading tasks that's surprisingly simple to do. And if you keep bees, you can use the beeswax from your hive in your handmade lotion!

Here's what else you need to make lotion:

- Double boiler (or a heat-proof bowl in a medium pot)
- Small pot
- Medium mixing bowl
- Larger bowl or baking dish
- Ice cubes
- Immersion blender
- Spatula
- Jar or container for storing lotion

None of the supplies should be made of aluminum.

SMALL STEPS

A double boiler works by putting water in the bottom chamber and the substance to be heated in the top chamber. The stuff in the top chamber melts from the heat of the boiling water, but it doesn't scorch or burn because it's not sitting on direct heat.

Three-Ingredient Lotion

At its most basic, lotion consists of just three elements: oil, water, and a binder. The binding substance is called an emulsifier. Without it, the lotion would separate into two layers, like homemade salad dressing does. The emulsifiers make everything creamy and lotion-y. Beeswax is a good emulsifier.

What follows is a basic recipe for lotion using these three ingredients. This simple, natural lotion can be made in a small kitchen and only takes about 20 minutes. The ratios in this recipe aren't set in stone, so feel free to adjust and create the type of lotion you're looking for. Increasing the amount of oil results in a thicker lotion; increasing the water gives you a lighter lotion. It's important to use distilled water, instead of tap water, when making lotion.

> $^{1}/_{2}$ cup olive oil
>
> 2 TB. beeswax
>
> $1^{1}/_{2}$ cups distilled water
>
> a dozen or so ice cubes

1. Put the oil and the beeswax in the double boiler, set it over medium-low heat, and warm until the beeswax is completely melted.

2. While the oil is warming, put the water into the small pot and heat on high until simmering (but not a rolling boil).

3. Place the ice cubes into the large bowl. Pour the warm oil/beeswax mixture into the mixing bowl, and set that bowl on top of the ice. Begin blending the oil mixture with the immersion blender.

4. Slowly, in a thin stream, pour the warm water into the oil mixture as the immersion blender is running. Continue blending until the mixture is thick.

5. Scrape the lotion into the jar, and store in the refrigerator. Anything that contains water (but no preserving chemicals) can develop bacteria, so it will extend the lotion's shelf life if it's kept in the refrigerator. Even when refrigerated, natural lotion will only keep for about a month, so it's good to make small batches.

Customizing Your Lotion

There are an infinite number of ways in which you could customize your lotion. One of the many benefits to lotion-making is that you can create something perfectly suited to your skin type and fragrance likes. You can create your own recipes, or use one of the many that are available on the Internet. Many of these items can be purchased at health food stores; others will need to be ordered online.

Oils: There are numerous oils you can use to make lotion, either alone or in combination, including avocado, coconut, cocoa butter, jojoba, macadamia nut, sesame seed, sunflower seed, almond, and grape seed.

Herbal infusions: Instead of using plain distilled water, try infusing it with herbs. Boil the water, pour it over dry herbs, and steep for an hour. Strain out the herbs, and make your lotion. Lavender flowers work great for this.

Other water ingredients: In place of a portion of your recipe's water, you can substitute aloe vera (good for inflamed or itchy skin) or vegetable glycerin (makes the lotion more hydrating).

Other emulsifiers: Small amounts of borax or xanthan gum can be added to the water to improve the lotion's texture. Vegetable-derived emulsifying wax is regarded as the best emulsifier for lotions, but it's not completely natural.

Essential oils: These can add both fragrance and skincare properties to your lotion. Use with restraint because they are potent—only about 50 drops per cup of lotion. Lavender, rose, and sandalwood are popular choices.

URBAN INFO

In Chapter 19, I note that you should wear gloves when using cleaning products that contain borax. However, when borax is used in lotion making, its skin-irritating properties are neutralized by the oil, similar to what happens to lye in soap. You should still wear gloves when creating the lotion, though—the borax isn't safe until it's neutralized.

Bonus oils: Adding a little of a specialty oil like evening primrose can boost the moisturizing power of your lotion.

Natural preservatives: Grapefruit seed extract, rosemary oil extract, or vitamin E can be added to ward off bacteria and increase the lotion's shelf life.

Spinning Homegrown Wool into Yarn

If you're raising angora rabbits or pygora goats, you're going to be harvesting angora wool fiber. If you'd like, you can use this wool to make yarn, which can then be knitted into socks or hats or any number of other things. However, you might not have room to bring a large spinning wheel into your urban home. Luckily, you can spin wool into yarn using a small hand tool called a drop spindle.

A top-whorl drop spindle can be used to spin fiber into yarn.
(Courtesy of BrianKraft.com)

Making a Drop Spindle

You can purchase simple drop spindles from craft stores for around $10, but you can make your own for even less money.

Here's what you need to make a drop spindle:

- CD
- 10-inch wooden dowel; should be able to fit snugly in the CD
- Small cup hook with a screw end
- Sandpaper
- Wood glue
- Small saw (optional)

The CD acts as a weight, which spinners call a *whorl*. The whorl can be placed on either the top or bottom of the spindle. Most of the instructions for home spinning use a top-whorl spindle.

Here's how to create your own:

1. Sand the dowel until smooth. Leave one end of the dowel completely flat—this will be the top of the spindle. The bottom end can be sanded down and rounded.

2. Put a dab of glue on the inside of the CD, and slide it over the top of the dowel. Position the CD on the dowel an inch or so from the top. Wipe away any excess glue, and let dry.

3. Screw the cup hook in the exact center of the top end of the dowel.

The Art of Spinning

Creating yarn from fiber involves spinning the drop spindle, which twists the attached fiber bit by bit into yarn. The length of the spindle collects the formed yarn.

There are different ways to position both the fiber and the spindle. It can be difficult to describe in a book; it's much easier to learn if you can watch someone do it. In Appendix B, I've listed the web address for a great instructional video on spinning with a drop spindle, and many more videos are available on YouTube.

SMALL STEPS

Angora fiber is soft and slippery and can be difficult for beginners to spin. Combining angora with sheep's wool that's been prepared for spinning (which is called roving and is available at specialty craft stores) makes it easier to spin.

The Least You Need to Know

- You can make soap, lotion, and yarn using materials produced on an urban homestead.
- Homemade body products are not only completely natural, but they're also more affordable than purchasing them at the store.
- Soap-making requires a few safety precautions but can be easily done in a city kitchen.
- You can spin fiber into yarn using a simple spindle you make yourself.

Cleaning Your Home, Naturally

In This Chapter

- What you need to make DIY cleaning products
- When to use bleach
- Recipes for cleaners, disinfectants, soaps, and more
- Natural cleaning tips

A trip to the cleaning-supply aisle of the grocery store might convince you that cleaning your home is a serious business, requiring dangerous-sounding chemicals packaged in endless combinations. In fact, you can clean your home safely and easily using products you make yourself.

Creating your own all-natural cleaning supplies is beneficial for a number of reasons. The chemicals used in commercial products are toxic to the environment. Plus, each type of product comes in its own packaging, which creates a lot of waste. Many people experience a range of health problems when they try to clean using commercial products, including skin and lung irritation. And last but not least, commercial cleaning products are expensive! For a relatively small amount of money, you can purchase the ingredients to make multiple batches of all kinds of cleaning products at home.

In this chapter, I give an overview of the ingredients and supplies you can use to make your own natural cleaning products. I provide you with several recipes and additional tips for cleaning your home. Plus, I talk about cleaning situations in which you may not want to limit yourself exclusively to natural products.

Gathering Your Ingredients

You can use a handful of simple, all-natural ingredients to create a wide variety of products for cleaning your home.

URBAN INFO

Read the labels on the cleaning products sold at the store. The main ingredient is almost always water. When you buy commercial cleaning products, you're paying to have something that's mostly water packaged and shipped to you across the country.

From Your Cupboard

Some of the most effective cleaning supplies come out of the kitchen. Check your cupboards—you might have most of these already on hand!

Baking soda: Even if you don't have a lot of storage space, it's worthwhile to buy the big box (or bag) of baking soda. It's a great deodorizer and can be used as a mildly abrasive scrub. When combined with vinegar, baking soda forms a fizzing mixture that has additional cleaning properties.

Vinegar: This can be used for cleaning, disinfecting, and deodorizing. It also works for cutting grease. Distilled white vinegar is usually the best (and cheapest) choice.

Lemon juice: This is good for cleaning, deodorizing, and cutting grease. Bottled lemon juice is more cost effective and stores better than fresh lemons.

Kosher salt: This salt is abrasive and used for scrubbing.

Cleaning Helpers

These natural products can help give your cleaning a boost:

Castile soap: A simple, mild soap traditionally made with olive oil, now the term is often applied to any soap made with vegetable fats instead of animal fats. Liquid castile soap makes a great base for a number of cleaning products. It can be purchased at health food stores, or you can make your own (see Appendix B).

Hydrogen peroxide: Available in the first-aid section of grocery or drug stores, it can be used as a disinfectant and also has bleaching properties.

Borax: A natural mineral compound, borax is a white powder that's available in the cleaning sections of grocery stores. It can be used as a cleanser and disinfectant.

Washing soda: Also known as sodium carbonate, it can be used as a water softener in detergents, removes grease, and deodorizes. It can be purchased at the grocery store.

ROAD BLOCK

It's a good idea to wear gloves if you'll be coming into contact with products that contain borax or washing soda. These ingredients should also be stored out of reach of children and away from pets.

Extra Power

Essential oils can be used to boost the disinfecting ability of natural cleaning products. Many essential oils also smell terrific, so they're natural deodorizers. The following essential oils are most frequently used in homemade cleaning products:

Tea tree: This oil has strong antibacterial qualities and is a great disinfectant.

Eucalyptus: Also a good disinfectant, this can be a deodorizer.

Lemon or grapefruit: Great for cleaning and deodorizing, these can be used in product recipes in place of lemon juice.

Lavender: This is a good deodorizer.

When Sterilizing Is Important

When you're cleaning, your goal is to eradicate all the germs, right? Actually … not always. When cleaning, it's good to remove dirt and grime, nasty odors, and a reasonable amount of bacteria. However, there's something to be said for living with a certain amount of germs. It helps our immune system grow strong. Some parents have found that children who grow up in an overly sterile environment are more susceptible to colds as compared to their peers who are allowed to play in the dirt. Routine use of natural cleaning products is a good way strike that balance of living in a clean, but not completely sterile, environment.

The natural disinfectants listed in this chapter can be counted on to take care of everyday germs, and using hot water when cleaning is also helpful. But there are times when eradicating bacteria and bacterial spores is more important, such as when members of the household are ill. Also, when pets have an "accident" involving feces indoors, it's necessary to take extra steps to be sure no dangerous bacteria are left behind.

These are situations in which you may want to consider using small amounts of bleach. Including bleach as part of your regular cleaning routine isn't advisable for a multitude of reasons. Mixing bleach with other cleaning agents can create toxic gas, it is irritating to people with sensitive skin, and chlorine buildup (from bleach) has an environmental impact—just to name a few. However, there are some occasions when the benefits of bleach outweigh the drawbacks.

SMALL STEPS

Sterilizing milking equipment is another situation in which using bleach is necessary. If you're raising backyard goats for milk, do not use the bleach alternative described in this chapter to soak your equipment—use real bleach water.

Supplies

Before you move ahead with making your own cleaning products, you'll want to have a few things on hand. The right supplies can make cleaning a snap!

Spray bottles: These can be purchased new in the cleaning-supply section of the grocery store, but empty bottles from commercial cleaning products work just as well—and sometimes better. Be sure to label your bottles, so you can keep your glass cleaner and all-purpose cleaner straight.

Shaker jars: Use these to sprinkle scrubbing ingredients like baking soda or kosher salt. Empty Parmesan cheese shakers are good, or you can purchase a new shaker jar from a kitchen store.

Cloth rags: If you'd like a large supply, visit a thrift store or garage sale and pick up some old T-shirts to cut into rags.

Sponge: This can be easier to use than rags, but it needs to be sterilized periodically. Heat a completely wet sponge in the microwave for two minutes to kill the germs.

Wool (old socks or a cut-up blanket): When wool is rubbed on a surface, it creates a static charge, which allows you to dust effectively without any kind of spray.

Newspapers: Rags won't work for cleaning glass or mirrors, and buying paper towels is unnecessary. Just crumple up yesterday's newspaper and use it to wipe away the glass cleaning solution—nothing works better!

Stiff cleaning brush: Sometimes the magic ingredient in natural cleaning products is "elbow grease." The products you create will be helpful, but at times, you just need to spend a bit of time scrubbing.

Bucket: Use this for mixing larger amounts of cleaning solutions. There's really no reason to purchase one—just find an old laundry soap or cat litter bucket.

Cleaning Product Recipes

Once you start making your own cleaning products, you'll be amazed that anyone pays inflated store prices for a bottle of water and chemicals when the homemade alternative costs just pennies and doesn't emit annoying fumes.

Creating cleaning products isn't like baking bread—there's no need to lock yourself into specific measurements. Use these recipes to get started, but feel free to experiment. None of the items covered in the "Gathering Your Ingredients" section earlier will cause a bad reaction if mixed, so you can improvise when creating your products. Remember, it's best to wear gloves if you'll be touching a product made with borax or washing soda.

All these recipes have been scaled down so you won't need a lot of space to store your cleaning products.

All-Purpose Cleaner

Here's what you need to make a good, general cleaner:

3 cups warm water

1 tsp. liquid castile soap

3 TB. white vinegar or lemon juice

1 tsp. borax (optional)

4 drops tea tree oil (optional)

4 drops eucalyptus oil (optional)

Pour the water into an empty spray bottle, and add the rest of the ingredients. If you put in the soap and then try to add the water, the mixture will foam out of the bottle. Cap the spray bottle, and shake gently to blend. Use the spray on any surface except mirrors and glass. Wipe with a sponge or rag to clean.

You can leave out the borax if you have sensitive skin and want to be able to use the cleaner without wearing gloves. If you'd like, you can keep it simple—the mixture will still work great with just water, soap, and vinegar.

Bleach Alternative/Disinfectant

Here's what you need to make a bleach-free disinfectant:

>   3 cups water
>
>   1½ TB. lemon juice
>
>   ¼ cup hydrogen peroxide
>
>   For an extra boost of disinfectant, add
>
>   10 drops tea tree oil

You can mix this in an old vinegar or bleach bottle if you want to use it for laundry, or mix it directly in a spray bottle to use as a disinfectant. Use it to wipe counters, cutting boards, or other surfaces you'd like to disinfect. Leave out the tea tree oil if you're using this as a bleach substitute for your laundry.

Liquid Dish Soap

Here's what you need to make dish soap:

>   2 cups liquid castile soap
>
>   15 drops lemon or grapefruit essential oil

Mix in an old soap bottle or glass jar, and use to hand-wash dishes. Do not use this product in an automatic dishwasher.

Powdered Dishwasher Soap

Here's what you need to make dishwasher soap:

>   1 cup borax
>
>   1 cup washing soda

Put 2 tablespoons of the mixture into the dishwasher soap dispenser for each load.

SMALL STEPS

If you're looking for a solution to the water spots that so often occur with automatic dishwashers, put distilled white vinegar into the dispenser that's meant for the clear rinse gel.

Laundry Soap

Here's what you need to make laundry soap:

12 cups water

¼ cup baking soda

¼ cup washing soda

¼ cup borax

½ cup liquid castile soap

10 drops essential oil for fragrance (optional)

Place the baking soda, washing soda, and borax in a bucket or large bottle (must be at least 1 gallon). Add 4 cups of hot tap water, and stir until the powder dissolves. Put the remaining 8 cups of water into the bucket. Add the soap and the essential oils. Stir. Let the soap mixture rest for 24 hours. It should be slightly gel-like. Use ½ cup per load of laundry.

Floor Cleaners

These recipes are a bit different, depending on whether you have hardwood or tile/vinyl flooring:

Here's what you need to make a hardwood cleaner:

4 cups water

½ cup white vinegar

10 drops lemon essential oil (optional)

Here's what you need to make a tile or vinyl cleaner:

4 cups water

½ cup liquid castile soap

10 drops lemon or eucalyptus essential oil (optional)

Mix the ingredients in a bucket, and use a rag or a mop to clean the floor. When finished, wipe the floor with a dry rag.

Glass Cleaner

Here's what you need to make glass cleaner:

$2^{1}\!/_{2}$ cups water

$^{1}\!/_{2}$ cup vinegar or lemon juice

4 drops lemon essential oil (optional)

Mix in a spray bottle. Spray on windows and mirrors, and wipe clean using crumpled newspaper.

SMALL STEPS

It's best not to clean your windows on a sunny day. The sun will dry the cleaning solution too quickly, which will cause streaks.

Natural Cleaning Tips

Having the right natural products and supplies makes your cleaning chores much easier. But sometimes success with cleaning is not only about what you use but how you do it.

Cleaning Your Toilet

Here's what you need to make your toilet bowl shine:

White vinegar

Baking soda

All-purpose cleaner

Sponge (or rags)

Toilet brush

Pour about a capful of white vinegar into your toilet bowl, and add a healthy sprinkle of baking soda. Allow them to steep in the bowl while you clean the rest of the toilet with the all-purpose cleaner. After that's done, sprinkle additional baking soda onto a toilet brush, and use it to scrub the inside of the bowl.

SMALL STEPS

It's wise to have a specific sponge you use only to clean your toilet. It's not a good idea to use the same sponge on the bathroom sink that you used to clean your toilet.

Unclogging Drains

Here's what you need to make a clog-busting solution:

$\frac{1}{2}$ cup baking soda

$\frac{1}{2}$ cup vinegar

Rubber stopper (or something else that can block the drain)

1 gal. boiling water

Pour the baking soda down the drain, and follow it with the vinegar. Quickly cover the drain with the stopper, so the fizzing action works to break up the clog. Wait 10 minutes, and pour the boiling water down the drain to flush everything away.

Scrubbing

Here's what you need to make a good scrubber:

Baking soda or kosher salt

Cleaning brush or sponge

Anything that's particularly stubborn—burned food in the bottom of a pot or built-up gunk on a counter—can be scoured away. Sprinkle on the baking soda or kosher salt. Wet the brush or sponge very lightly (too much water will cause the abrasive powder to dissolve), and scrub. If you're still having trouble, soak the area with baking soda and a little water for a few hours, and try scrubbing again.

Removing Stains

Here's what you need to make cleaners for various stains:

For hard surfaces (counters, tiles): Make a paste out of baking soda and water. Apply it to the stain, and let it stand for several minutes. Scrub with a cleaning brush.

For mildew (shower or bath): Spray it with vinegar or lemon juice. Wait a few minutes, and scrub with a cleaning brush.

For clothes: Make a paste out of washing soda and water. Apply it to the stain, and let it stand for a while. Wash the clothes. There's no need to wipe away the washing soda paste.

For carpets: If the spill is fresh, soak the area with club soda and blot away. For an older spill, try covering it with baking soda and then pouring white vinegar on top. As the mixture foams, scrub gently with a rag or cleaning brush. When the foaming stops, blot with a dry rag.

Eliminating Carpet Odors

This trick has been well publicized by the people who make baking soda: Sprinkle baking soda liberally over your carpet. Wait 30 minutes and then vacuum it up.

The Least You Need to Know

- You can clean every part of your home—from windows to floors—using natural, homemade products.
- A small number of simple ingredients can be used to create a variety of cleaning solutions.
- There are times when you may want to use bleach to kill especially nasty germs.
- You can learn a few tricks to make cleaning with natural products easier.

Making the Most of What You Have

Homesteading isn't always about creating something new. It's often about efficiently using the resources that are already available. This is where having the opportunity to homestead in the city is really a blessing—cities have an abundance of resources a country dweller could only dream of!

Whether you want to build using salvaged materials, compost to your heart's content, or forage for free food, you can find what you're looking for in the city. You can also harness the power of the sun and wind to help power your home, stretch your water farther, and discover lots of environmentally friendly ways to get around town.

In this part, I talk about how to make the abundance of resources available in the city work for you.

Energy-Wise Living

In This Chapter

- How we "spend" most of our energy
- Simple ideas for decreasing your energy use
- Options for getting off the grid
- Getting around town—sustainably

Have you ever been forced to go without electricity for a long time due to an extended power outage? There's nothing like a forced period of low-tech living to show us how much energy is woven into our daily lives—and how much we take it for granted. It can't be denied that our modern society uses tremendous amounts of energy, both within our homes and for transportation.

However, simply replacing our current dependence on resource-depleting energy sources with equal amounts of sustainable energy isn't a complete solution. We would do well to first look at how we can decrease the volume of energy we need to get through the day and then start to look for sustainable sources.

In this chapter, I begin by discussing ways to lessen our energy needs and make better use of what we have. Next, I talk about day-to-day activities you can accomplish without using any energy. I cover options for generating your own energy from the sun and also look at ways you can reduce your energy usage through alternative transportation.

How We Use Energy

Before we can examine ways to save and generate energy, we need to understand how we use energy in our home. To many people, the most obvious consumers of energy are all the appliances we plug in and turn on. Or it may seem to be the lights we're constantly turning—and leaving—on.

Actually, the biggest household energy drain has to do with something entirely different—temperature. The largest chunk of energy usage by far, about half a home's total use, comes from heating and cooling the air.

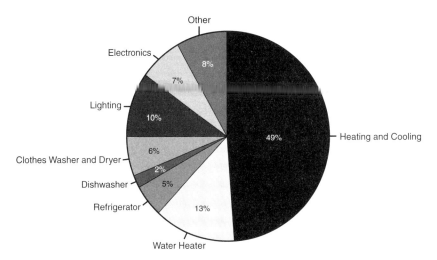

A breakdown of average home energy usage, according to a Residential Energy Consumption Survey. "Other" refers to household appliances like stoves, ovens, microwaves, and coffee grinders.

No other category of energy usage even comes close to what we spend moderating our home's temperature. The next largest individual drain is heating our water, followed by lighting. The appliances we use to feed ourselves and keep our clothes clean total up to 21 percent, and electronics like TVs or stereos use 7 percent.

Each of these categories of energy usage presents opportunities for conservation—and even doing some things without energy altogether. In the case of entertainment-based electronics, the solution is often to simply turn them off and do something else. This is when being an urban homesteader is helpful because you have the option of milking the goats or making lotion instead of watching television!

Pretty much all the advice related to saving energy in traditional homesteading books assumes that the homesteader lives in a house—a house upon which you can mount solar panels or install various alternative heaters, plus a yard to hold additional energy projects. Sometimes they even focus a lot of attention on how to position one's house (with the idea, apparently, that the homesteader will get a piece of land and build from the ground up). It's certainly possible for some urban homesteaders to do these things. However, if you're renting an apartment, you often don't have the space—or permission for construction projects!—that may be required. Nevertheless, there are several small but powerful things you can do to save and use energy more sustainably.

Powering Down

A penny saved is a penny earned.

—Benjamin Franklin

Or to paraphrase Mr. Franklin, a *kilowatt-hour* saved is a kilowatt-hour earned. Even the most gung-ho sustainable energy advocates agree it doesn't make sense to run out and buy solar panels or a wind turbine without taking steps to reduce the amount of energy you're using in the first place. Figuring out how to cut your energy usage in half is as good as developing a way to generate half of your energy—in fact, it's better. The best part is that these solutions cost no or very little money.

DEFINITION

A **kilowatt-hour** is a unit of energy equivalent to 1,000 watts of power expended for one hour. Kilowatt-hours (kWh) are the way household energy use is measured and billed by the utility company.

Heating and Cooling Your Home

Throughout this book, I present the many advantages urban homesteaders may have over their rural counterparts. Finding ways to use less energy to control the temperature of your home is no exception. In fact, many urban dwellers are on the cusp of a growing trend without even realizing it!

Over recent decades, the size of the average American home has increased substantially—and with it, there's been a significant increase in the cost of heating, cooling, and maintaining those homes. In response, there's rapidly developing interest in what is known as "small space living" or the "tiny house" movement. Architects and sustainable living experts are designing houses as small as 100 square feet. (See Appendix B for a video link that will take you on a tour of one of these homes.)

While you certainly don't have to go that far, there is something to be said for choosing to live in a smaller space—even if you can afford something larger. Our satisfaction in life is largely due to our perspective. If you've been feeling down-and-out because your living space is less than palatial, instead try to look at it as the very latest in sustainable living. Your 400-square-foot studio apartment is trendier—at least within the "green" community—than you thought!

SMALL STEPS

In terms of energy consumption, be especially wary of any living space with high ceilings. Trying to heat and cool those rooms is typically inefficient.

Before you can plug your energy leaks, you need to know where to begin. Most local energy companies offer a free or discounted energy audit to let you know your home's biggest problem areas. Some nonprofit agencies also sponsor free energy audits for qualifying residents.

The energy company dutifully generates power to heat and cool your home, and you faithfully pay the bill, but a good chunk of that energy is just leaking away. This is probably the biggest opportunity for energy savings in most homes. Getting your home weatherized is an investment, but if you're planning on staying in the same place for a few years, it'll pay for itself in decreased energy bills. Plus, it's likely that you can qualify for a tax credit to help cover the cost. If you're not sure how to start, the energy auditing folks will be more than happy to give you ideas. If you can only do one thing, consider replacing your windows with an energy-efficient variety.

Set the thermostat lower in the winter. This may seem like an obvious part of the solution, but you might be surprised at how effective it can be. For every 1°F you lower your thermostat, you can reduce your energy use by approximately 3 percent. Experiment with finding a new baseline for your home during the winter months.

Another relatively low-tech option is to use a programmable thermostat, which automatically decreases the temperature in your home during the day while you're at work or at night while you're sleeping.

Choose fans instead of air conditioning. Fans aren't a no-energy solution, but they certainly use fewer kilowatt-hours than an air conditioner. Ceiling fans can be great at moving air, and even small portable fans are helpful. If you live in a free-standing home, you can also consider installing a whole house fan (also called an attic fan). These aren't great to use when it's especially hot outside because they will draw the hot air in. However, they're helpful in the evenings when the outside temperature has dropped but the inside of the house may still be warm.

If you're renting an apartment and aren't able to make permanent insulation fixes (or if you live in a house and are in the same situation), you can affix clear plastic over your windows in the winter to stop heat loss. Hardware stores sell large rolls of plastic meant for this purpose (you use a hair dryer to shrink-wrap them to the window frames), or you can cut sheets of bubble wrap and attach them over your windows with a little tape.

Clean your furnace filter. A filter that's full of gunk causes your furnace to work much harder—and use more energy—to produce heat.

Water Heating

You can save a good amount of energy over time by lowering the base temperature on your water heater. Most people don't like the water that comes out of the shower or faucet to be scorching hot anyway—they end up diluting the hot water by turning on the cold spigot. Bring your hot water down to a comfortable temperature and lower your energy usage in the process!

A chunk of the energy your water heater creates is lost to the surrounding air. Wrapping your heater in an insulating blanket helps redirect the energy back toward its intended use.

Lighting

You can save a lot of energy—and get a longer-lasting bulb—by switching to compact fluorescent lights (CFLs). Each bulb costs a little more than a standard incandescent light bulb, but it lasts eight times longer and uses one quarter of the energy. According to the U.S. Energy Information Administration, widespread use of CFLs could decrease overall household energy usage by 7 percent.

ROAD BLOCK

CFLs contain mercury. The bulbs aren't dangerous under normal circumstances, but if they're broken, they can leak. A broken bulb doesn't require bringing in a professional Hazmat team, but it should be handled with care. Go to epa.gov/cfl/cflcleanup.html to read the EPA's recommendations for cleaning up a broken CFL bulb.

Appliances and Electronics

Even if you turn off your appliances and electronics when you're not using them, they still may be drawing electricity. Anything with a clock (like your microwave) is notorious for doing this. This constant energy drain is called the "phantom load"—the load certain gadgets consistently put on the energy supply. An easy way to address this is to plug all your appliances and electronics into a power strip and then flip the switch on the power strip when you're not using the appliances. You can also purchase special power strips that automatically cut the phantom load.

When it comes time to purchase a new household appliance, consider buying models that have received an Energy Star rating for efficiency. They do cost a little more than the non-energy-efficient models, but you'll make the difference back over time in lowered energy bills. Tax credits are available when you purchase an Energy Star appliance.

Speaking of Energy Star appliances, if there's only one thing you can upgrade, make it your refrigerator. That single appliance accounts for 5 percent of your total household energy use. It's also worth taking an honest look at what you keep in your refrigerator and whether you (and the environment) might be better served by downgrading to a smaller model. Shay Salomon, author of *Little House on a Small Planet*, says most refrigerator space is taken up by C&C—condiments and compost.

Keep your appliances maintained and in good repair. Poorly functioning appliances needlessly drain away power. One good tip is to periodically wipe the dust off the coils on the back of your refrigerator. Fridges keep things cold by discharging warmth through the coils, and if they're clogged, the fridge has to draw more energy to keep cool.

Many dishwashers offer a "low energy" option, during which the dishes air-dry in the machine instead of being forcibly dried with heat.

URBAN INFO

It's logical to assume that if you want to save energy, not to mention water, you'd be better off hand-washing your dishes instead of using the dishwasher. Surprisingly, a study done by the University of Bonn says that dishwashers use ½ the energy and ⅙ the water (less soap, too!) compared to hand-washing. So feel free to devote your time to other homesteading tasks and let the dishwasher do the work!

Also, pay attention to how long you ask your clothes dryer to run. It may not need to go for the maximum amount of time to get your clothes dry.

Freedom from Energy

Once you've taken steps to "power down" your home as much as possible, you can start looking at ways to accomplish some of your daily living activities without using any energy at all. Sometimes these changes involve working with what's naturally available, and sometimes you may have to make small, commonsense behavior changes. You could also learn some new household skills, which are actually old tricks homesteaders have been using for years.

Heating and Cooling Your Home

Adjusting your life to limit or do without energy-consuming heating and cooling requires a shift from the modern-day way of approaching comfort. It's not unreasonable to look at twenty-first-century life and conclude that we do everything possible to disconnect ourselves from seasonal changes. Most people expect to be able to always wear short-sleeve shirts indoors, even during the dead of winter. On the other end of the spectrum, sometimes the air conditioning is turned up so high during the summer things become chilly indoors.

Although there's certainly a case to be made for supplemental heat and cooling in some situations, we'd do well to remember that central heat and air conditioning have only been in widespread use for a few decades. Somehow our ancestors managed without them!

URBAN INFO

AC-free advocates state that going without air conditioning can actually be easier on your body in the long run. As the temperatures rise, your body acclimatizes. However, if you're constantly going between the chill of an air-conditioned space and the heat of the outside, it can cause physiological stress. (Those who are vulnerable to heat stroke, like the very young and the elderly, can be better served by having air conditioning.)

Regulate from the inside out. As simple as it sounds, drinking hot beverages in the winter and cool beverages in the summer goes a long way toward making seasonal temperatures bearable. Some people find that adding a little raw apple cider vinegar to their drinking water during the summer helps their body deal with the heat more effectively. Also, a surefire way to raise your body temperature in the winter is to exercise. It's a perfect way to stay in shape when you can't work in the garden!

Dressing appropriately is smart, too. Be honest—have you ever turned up the thermostat instead of reaching for a sweater? Dressing in layers during cold months can do wonders for keeping warm. In the warm months, remember that cotton fabrics breathe more than polyester. While you're inside, you can dress as skimpily as you'd like (bathing suits, anyone?), but when you go outdoors, opt for loose-fitting clothes that cover your skin—a sunburn will jack up your body's temperature and make the heat more unbearable. If you'd like a personalized evaporative cooler, try wearing a wet long-sleeved cotton shirt.

SMALL STEPS

Plant deciduous trees, the kind that lose their leaves in the winter, in front of your south-facing windows. In the summer, their foliage blocks your windows from the heat of the sun, and in the winter, their bare limbs let the sun shine through. You can also plant evergreen trees on the north side of your home, to protect from winter winds.

Your windows can do a lot to regulate the temperature in your home. If you'd like to trap heat in the winter, be sure the blinds and curtains covering south-facing windows are open during the day. During the summer months, south- and west-facing windows should stay covered. If you feel your home has become too dark, use a white sheet to cover the windows.

You can create some strategic airflow by opening the windows on the north and east sides of your home and just barely cracking the west-side windows. As an additional bonus, create your own natural evaporative cooler by hanging wet sheets over open windows (with a towel underneath to catch the drips). At night, when the outside temperatures cool off, open as many windows as you can to release any heat trapped in your home.

ROAD BLOCK

Leaving windows or doors open to facilitate airflow can present obvious safety risks in an urban environment. Some people choose to protect themselves by installing iron security bars over their doors and windows. While this presents obvious aesthetic drawbacks, some homesteaders find the energy-saving benefits worth it.

Much of the sun's heat that affects your home comes through the roof. Unfortunately, this is something you can't switch out with the seasons, but you can paint your roof to affect whether it absorbs or repels the sun's heat. If you suffer from hot summers, white paint will deflect the sun's rays. If cold winters are more of a problem, paint the roof black so it will better absorb heat.

When doing your landscaping, locate the plants that need the most watering close to your house. The natural evaporation of that water from the plants' leaves (and the soil) will help to keep your home cool.

If you can create some kind of an outdoor living space—whether it's a balcony, a screened-in porch, or a patio—on the north or east side of your home, it will be a cooler place to spend your evenings. As a bonus, some families find the need to escape the house during the evenings provides a good opportunity to spend time together.

Wood-burning stoves and fireplaces are perhaps easier for rural homesteaders with access to tree-filled land to utilize, although they can also work in a city. The challenge is to find a reliable source of wood. Scrap lumber can be used, but you'll want to be sure it hasn't been pressure-treated with a bunch of nasty chemicals before you burn it in your home. A good option would be to make friends with someone who works for a tree-trimming company and arrange to periodically collect their cut logs. Check with your local fire department to make sure that wood stoves are legal where you live.

If you have a vent on the south side of your home, you can build a box that will collect the sun's heat and direct it inside. The efficacy depends, of course, on the amount of sunshine your area gets during the winter months. See Appendix B for websites with instructions for building a solar space heater.

Prevent any heat you create in your home—whether it's intentionally through a fireplace or space heater or secondarily as a result of cooking—from dissipating as much as possible. Keep your doors closed to trap the heat in the areas where you're spending the most time.

Water Heating

A wide variety of options are available for creating a solar water heater, ranging from relatively simple do-it-yourself projects to extensive systems that need to be professionally installed. On the simpler end is something called a batch (or breadbox) system. It basically involves pumping water into a black tank stored inside an insulating glass box. The heated water can be either pumped directly out of the tank or stored in a separate storage tank to prevent it from mixing with the still-being-heated water. Luckily, plans are available online—see Appendix B for a website.

ROAD BLOCK

Unless you construct a closed-loop system that incorporates antifreezing chemicals, you'll need to drain your batch solar water system in the winter to prevent the pipes from freezing.

If you'd like to start with a slightly less ambitious project, consider fashioning an outdoor solar shower. We all know how quickly water can heat up if a full garden hose is left in the sun (especially if it's a black hose). Solar showers take advantage of this phenomenon, either by using long coils of hose that are filled with water and left to warm or by filling a black bag (with a hose attached) with water and leaving it out in the sun. The hose and/or the bag are suspended overhead, and gravity causes the water to flow out the nozzle at the end of the hose. Some solar showers incorporate a Y connector so the user can temper the hot water with cool water from another hose. Several solar shower designs are available online.

Lighting

A light tube is exactly what it sounds like—a tube containing highly reflective material that brings light from the outside to the inside. It usually consists of a small dome on the roof of a building, connected to another small dome (the "light") inside the building. Light tubes allow for better insulation and more flexibility than a skylight, but of course, they share the limitation of only being useful during the day.

Kerosene lamps were a great source of light a hundred years ago (much better than candles), and they're still good today. However, any flame-based light carries a significant risk of fire, so the potentially serious downsides may outweigh the energy-saving benefits.

SMALL STEPS

Don't be afraid to think outside the box when dreaming up ways to decrease your energy use. For example, why not get a little headlamp (the kind hikers use) and have some "lights off" nights? The headlamp's small battery surely uses less energy than a houseful of lights!

Appliances

Have you heard the story of the guy who stockpiled canned food in his pantry in case of an emergency and then went hungry when the power went out because he couldn't operate his electric can opener? A silly story, to be sure, but it exemplifies the ridiculousness of our reliance on energy-dependent gadgets to do what we could accomplish ourselves with a little elbow grease. Electric can openers and mixers are just the beginning. If you're so inclined, you can find a number of "unplugged" appliances that use a hand crank, including food processors, blenders, and coffee grinders.

A "solar clothes dryer" sounds fancy, right? Actually, it's just using a simple clothesline to dry your laundry. About as low-cost and low-tech as you can get, clotheslines are still a great way to save energy and money. If you're short on space, you can find a rotating collapsible setup.

URBAN INFO

As benign as clotheslines may seem, many areas across the country don't allow them. Sometimes they're outlawed by zoning codes, and sometimes HOA regulations forbid them. An organization called Project Laundry List has launched a "Right to Dry" campaign, aimed at helping people combat anti-clothesline laws. Go to laundrylist.org for more information.

Also known as solar cookers, solar ovens can be a great way to prepare food in the summer without heating up your kitchen. You can make a solar cooker with just some cardboard and tinfoil, but the most effective kind uses the same principles as a gardening cold frame or a solar water heater—an insulated box with a glass or plastic frame to amplify the sun's heat. As with most things solar, you can find plans online for building your own—see Appendix B.

Save a little energy by discontinuing the use of your garbage disposal. Give all those food scraps to your backyard animals or compost pile!

Getting Off the Grid

The ultimate in energy self-sufficiency is the prospect of generating your own power. Of course, technically speaking, energy can't be created or destroyed—the dedicated home-steader is simply converting energy from another source into usable electricity.

Although wind power is a developing technology, it's generally agreed that it's not a practical option for city dwellers. The very definition of a city means a mass of buildings, which are going to block wind and prevent any significant or consistent airflow. Plus, it's not a good idea to attach wind turbines to buildings because of the vibrations and noise they produce. The turbines would have to be hoisted onto poles at least 30 feet above the highest building, and it's not likely city planners are going to welcome that idea any time soon.

Solar systems using photovoltaic (PV) panels are the most practical option for urban homesteaders wishing to achieve more energy independence. However, these systems aren't cheap, even with all the tax credits available. They also require toxic chemicals and rare earth metals to manufacture. Therefore, it's wise to focus first on decreasing your household energy usage.

Once you get your monthly kilowatt-hours down as much as possible, you'll have a better idea of the solar system you'd need to support your usage.

Solar Panel Options (Even for Renters)

Utilizing solar panels for energy doesn't have to be an all-or-nothing proposition. You can choose to invest in enough panels to support all your household's energy needs, or you can use a couple of panels to supplement the municipal energy supply and build from there.

One especially important element for urban residents is the possibility of investing in a self-contained solar system. Self-contained systems consist of panels and electronics mounted together in a box, so you can take it with you when you move. Solar panels pay for themselves in energy savings, but it takes time. Self-contained systems offer the same benefits as house-mounted systems but with the perk of portability.

ROAD BLOCK

If you're going to invest in solar panels (especially the kind that are permanently mounted on your house), it's a good idea to talk with your neighbors. Be sure they won't be planting any big trees that will someday shade your panels!

There are two basic ways to set up a house-mounted or self-contained solar panel system—off-grid or grid-tied. A true off-grid system requires an extensive battery backup, which allows you to store excess power that can be used at night or when it's cloudy. Obviously, choosing an off-grid system means you'd have to invest in enough solar panels to provide all your home's energy.

If you'd like to take less of a full-scale approach, you can choose a grid-tied system. You stay connected to your city's electrical system, and you can give back to the grid—and receive a credit on your electric bill—when your solar panels produce more energy than you use. If your panels aren't producing enough energy, the city system will supplement as needed.

The main disadvantage to a grid-tied system is that, when the city's power goes down, your solar panels also stop working. This is called anti-islanding, and it's meant to protect power company employees from being electrocuted when they tap into the system to make repairs.

Solar-Powered Devices

You can harness the sun's energy without full-size solar panels. An ever-increasing number of small, solar-powered devices are available for you to use, both to decrease your energy usage and to achieve greater self-sufficiency. Some of the more popular choices include battery and cell phone chargers, radios, portable electronic battery packs (that can power things like laptop computers), flashlights, and car battery chargers.

Getting Around Town

Transportation isn't fundamentally a homesteading skill, of course, because it happens away from the home. However, it's most certainly a sustainable living issue. Transportation is the second-biggest source of energy consumption in the United States (second to electric power), according to the U.S. Department of Energy. It doesn't make much sense to take steps toward living sustainably at home through urban homesteading without making correlating efforts at sustainable transportation.

The good news is that—once again—urban homesteaders have a wealth of resources available to them. Most cities have continuously developing public transit systems, which are a surefire way to reduce your energy footprint. There are a multitude of other options for sustainable transportation within cities, and I discuss a few of them in the following sections.

Telecommuting

Sometimes the best way to travel sustainably is not to travel at all. Many companies—both large and small—are exploring the concept of allowing their employees to telecommute (work from home). While it isn't possible to telecommute if you have a direct service job, many computer-based occupations can just as easily be done from a home office. If you're interested in trying to implement a telecommuting program where you work, understand that it will likely be a lengthy process. However, you can start the conversation with your boss by sharing the following benefits of telecommuting:

- Improves employee job satisfaction, motivation, and morale
- Benefits the community by improving air quality and reducing traffic congestion
- Allows the employer to increase staff without increasing office or parking space
- Decreases employee stress and improves efficiency
- Serves as an effective, low-cost recruitment tool and addition to employee benefit packages

… and much more! To learn more, just do an Internet search for "telecommuting benefits." There is plenty of information out there.

Car-Sharing

Car-sharing is a transportation option unique to urban residents. (It's not currently available in most suburban and rural areas.) Car-sharing allows participants to arrange for short-term use of a car without shouldering the expense and responsibilities that go along with car ownership. The cars are insured and maintained by the companies, which also keep them clean and filled with gas.

A typical membership costs around $50 per month, plus $9 an hour to use a car. In comparison, the Automobile Association of America estimates that the yearly operating cost for a mid-size sedan, driven 10,000 miles, is $6,242 per year. If you add on the costs of parking (which can be considerable in some cities), drivers can expect to pay total monthly costs of up to $1,000. Compared to this, car-sharing is a heck of deal, without the hassle of scheduling oil changes or the headaches that come with car repair.

A couple of car-sharing companies operate nationally, and many more offer services on a local level. Search online for "car share (your city)" to see what's available in your area.

Making Your Bike Your Primary Vehicle

When it comes right down to it, the main reason (other than nasty weather) people choose to use their car instead of their bike for local errands is because they need storage space. It's hard to balance three bags of groceries, a box of printer cartridges, and a bag of dog food all on your bike's handlebars. However, you shouldn't let this challenge stop you from using your bike to take care of business around town. A few simple modifications—which certainly cost less than a few car payments—can upgrade your bike so you can use it as your primary means of transportation.

URBAN INFO

Many cities, including Chicago, Denver, and Washington, D.C., now also have bike-sharing programs. Participants can pick up a bike from one of the kiosks scattered throughout the city and ride it to their destination, where they park the bike at another kiosk.

One way to add storage space is by purchasing some type of bike attachment. These are available as kits that connect to the back of your bike, and some also extend the rear wheel base. A number of styles are available—everything from symmetrical cargo bags to a trailer that can hold gardening equipment. These kinds of bikes are often called Xtracycles (which is also the name of a company that makes them) or Xtrabikes.

Another option is to get a Dutch-style front storage unit. Some of these are as small as baskets, but they can be big enough to hold a couple of children. While extremely popular in the Netherlands, they aren't yet in wide use in the United States. They can be special ordered and shipped to you—see Appendix B for a resource.

This bicycle has been adapted with a Dutch-style front storage unit.
(Courtesy of BrianKraft.com)

Biodiesel

If the prevalence of deep-fried food in the American diet has any upside, it's that used vegetable oil is widely available in cities. If you're able to make a connection with a restaurant to haul away its used oil, you can use it to make biodiesel. For this to work, your vehicle must have a diesel engine.

SMALL STEPS

Those who make their own biodiesel often find that the oil from locally owned restaurants is of a higher quality than that from fast-food chains. This is thought to be because fast-food restaurants typically use (or reuse) their oil for much longer before discarding it.

To convert vegetable oil into well-functioning biodiesel, you need to process the oil to remove the sugars and starches. (Technically, you could pour the oil straight into your vehicle, but this is not the recommended method.) The process requires a way to heat the oil, a thermometer, a pH-testing device, methanol, and lye. Great instructions for making biodiesel (complete with photos) are available in *The Self-Sufficient Home: Going Green and Saving Money* by Christopher Nyerges (Stackpole Books, 2009).

The Least You Need to Know

- The largest percentage of home energy usage goes to heating and cooling the home.
- Before taking steps to generate your own energy, do what you can to trim your home's energy use.
- A number of household activities can be accomplished without energy.
- Urban residents have a variety of sustainable transportation options available to them.

Water Is Precious

In This Chapter

* What your household water is used for
* Simple ways to conserve
* Capturing rain
* How to reuse water

Many people view water as our most valuable resource. In developing countries, access to clean water is a key element in the health of people and communities. Some are calling water "the new oil" because they believe that, as it becomes more scarce, there will be ever-increasing conflict around it. Even if we step aside from larger environmental concerns, every homesteader who grows food knows how important water is—and how pricey it can be. Therefore, it makes sense to learn ways to better utilize available water as part of a sustainable lifestyle.

Whether you live in a house or an apartment, there are things you can do to improve the way you use water. In this chapter, I begin by looking at the ways in which we use water, so you can get a sense of the place (or places) where you'd like to direct your efforts. Then I talk about methods for conserving water, capturing it when it falls from the sky, and reusing what you can.

How We Use Our Water

Where does water fit into your day-to-day life? It's likely that your first thoughts are about the water you drink and then perhaps what you use to cook. Of course, we all use water for cleaning ourselves, our clothes, and our dishes. Toilets factor into the equation, too. All in all, a fair amount of water is utilized inside our homes.

However, the average American is responsible for 80 to 100 gallons per day of household water usage. Even if you take long showers, that still sounds like a heck of a lot of water. So where's it all going?

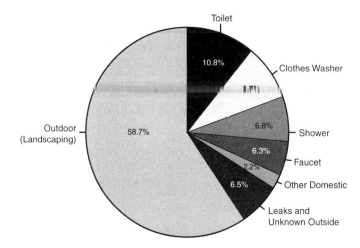

This residential water use summary shows where your water goes.
(American Water Works Association Research Foundation)

Actually, about two thirds (65.2 percent) of those 80 to 100 gallons are being used outside the home, and almost all that outside water is going to landscaping. In pretty much every case, "landscaping" means lawns.

Urban residents who live in apartments may feel they're off the hook in this particular area, and it's true that when it comes to landscaping, apartment living is a much more sustainable lifestyle. However, if your apartment building has a landscaped courtyard or shared grounds, that's going to count toward your water usage. Even if you can boast a total lack of water-consuming sod on the property where you live, it's still worth noticing all the grass that's being watered (for you, ostensibly) on city-owned land. If you're interested in working to change this practice, go to eatwhereUlive.com to learn what you can do about reducing water usage on your local public lands.

As you can see, the other 34.8 percent of residential water usage happens inside the home. Toilets take about a third of that, followed closely by clothes washers and showers. The "other domestic" category in the chart includes baths, dishwashers, and unspecified domestic uses.

Getting By with Less

As with energy, any steps toward sustainable water usage need to begin with conservation. Unlike electricity, however, we can't easily figure out a way to generate more water, so we need to make the best use of what we have. The following sections explore how you can conserve water both inside and outside your home. Even if the savings seem small—a drop here, a gallon there—water experts agree that every little bit helps.

URBAN INFO

Excessive water use isn't only about water resources but energy resources as well. According to the Center for Sustainable Systems, 4 percent of the United States' electricity goes toward moving and treating water and wastewater.

Inside Conservation Tips

Whenever you have the opportunity to replace or upgrade appliances and plumbing fixtures, opt for a low-water model. Front-loading clothes washing machines use 40 percent less water (and 50 percent less energy) than top-loading machines. You can also look for Energy Star models, which are created to decrease both water and energy usage—and usually get you a tax rebate. Purchasing a low-flow showerhead is a relatively inexpensive way to decrease your water usage, too.

One toilet flush uses more clean water than many people in underdeveloped countries have for their entire day's drinking, cooking, and cleaning. You can find toilets with a "dual flush" option, which allows the user to select from a 0.8- or 1.6-gallon flush so you can save water. If you're not at the point of upgrade yet, it's easy to convert an older toilet so it uses less water. Just fill a plastic bottle with rocks and put it in your toilet tank. This displaces some of the water, so less is needed to fill the tank and flush the toilet.

URBAN INFO

A few (quite dedicated) people are interested in using toilets that don't require water—even in an urban setting. Typically called "composting toilets," these can be purchased or made from scratch. There are some obvious health concerns associated with the composting and disposal of human waste, but these folks are exploring how to keep it safe. Go to humanurehandbook.com for more information.

You could be doing everything you can think of to conserve water throughout your home, but if something is leaking, then a lot of water is still going down the drain. A pipe or faucet that leaks 1 drop per second wastes 225 gallons of water each month. A good way to determine if you have a leak is to check your water meter, go three hours without using water, and check the meter again. Any movement in the meter means there's a leak somewhere in your system. A trick for sleuthing out leaks in your toilet is to put a few drops of food coloring into the tank. Wait 30 minutes and see if any of the color leaks into the bowl.

You're more likely to keep dripping faucets and running toilets in check if you have a few plumbing skills under your belt. You can find an abundance of how-to videos on YouTube, or check out the toilet repair information at toiletology.com.

Outside Conservation Tips

Get rid of your lawn. Really. There are very few definitive "you *should* do this" moments in this book, but there's no getting around the fact that lawns are an incredibly wasteful use of resources. If you'd like a green groundcover that uses less water and requires less maintenance than grass, try seeding with clover. Or if you're really attached to the idea of grass, switch to a native variety like buffalo or fescue. Even better would be to *xeriscape* your space or use it to grow food.

DEFINITION

Xeriscaping is landscaping in ways that reduce or eliminate the need for supplemental irrigation.

Xeriscaping typically utilizes plants that are native to the area, although it may also incorporate gravel and other nonliving mediums. You can find information on plants native to your area by doing an Internet search for "native plants xeriscape (your state)." Although it may need a little bit of watering to get established, the goal with a xeriscaped yard is to have it maintain itself using only natural precipitation.

You could plant food instead of grass. It's far more sustainable to use your yard space— whether it's in front of or behind your home—to raise food instead of inedible grass. From a homesteader's point of view, if you're going to be watering it, you should be able to eat it!

Raised beds may be all the rage in gardening circles nowadays, but unless you live in a particularly wet climate, doing the opposite makes more sense when it comes to water conservation. Concave or trench garden beds collect water flow and decrease the need for supplemental watering.

The practice of permaculture (which I discussed briefly in Chapter 5) involves observing your yard and noticing where water naturally flows and pools. If you'd like, you can exaggerate (or create) this effect by digging trenches, creating hills, and forming little channels through which water can flow. Put your water-loving plants, including most vegetables, in the damper parts of your yard, and place drought-tolerant plants in areas that stay drier.

Consider setting up a drip irrigation system. Drip irrigation uses about half the water of a standard sprinkler system. It can also help keep your plants healthier because they won't experience wet leaves, which can speed up the spread of disease.

Mulch, mulch, mulch. Placing wood chips, leaves, straw, or other organic matter around the base of your plants—or over any exposed dirt—significantly helps reduce water evaporation from the soil.

Rainwater Harvesting

Much of the water that falls from the sky goes to waste, at least from a homesteader's perspective. Storm water runoff goes into the sewer system, where it has to be processed and transported—and paid for—before you can use it to water your garden. Rain that falls in the city may contain lead above the limit recommended for consumption, and it will pick up debris and bacteria when it hits building surfaces, so harvested rainwater is not good for drinking. However, it can be a great way to quench the thirsty beast that constitutes the majority of our household water usage—outside irrigation.

If you happen to live in an apartment, don't assume the information in this section can't be used in your situation. Rainwater harvesting is pretty much always about roofs, especially in urban settings. Rain lands on the roof and travels through a gutter into a downspout. Apartment buildings have gutters and downspouts, too. It may take a little more creativity—and a conversation or two with the landlord—but you can capture some of that water. The book *Toolbox for Sustainable City Living* by Scott Kellogg and Stacy Pettigrew (South End Press, 2008) has some great models for rainwater collection.

Using Rain Barrels

Barrels are a popular receptacle for catching rain. Technically the barrels can be large or small, but many people like to make them out of old 55-gallon drums. You can purchase a premade rain barrel that's all ready to go, but adapting an old barrel is a fairly simple task.

Of course, whenever you repurpose an old container, you need to be sure it wasn't previously used to store something toxic.

Leaving the barrel uncovered allows mosquitoes to breed, so it's best to cover the top and just cut out a hole for the downspout.

If you're worried about roof debris in your water, use a piece of cotton fabric as a filter over the downspout hole.

Your barrel will also need a fitting for a hose so you can easily draw the water out of it. This can be done with regular PVC fittings—see Appendix B for a website that gives instructions.

 URBAN INFO

One gallon of water weighs 8 pounds, so a typical 55-gallon rain barrel weighs 440 pounds when full!

It's also a good idea to install an overflow pipe in the event the barrel fills. Don't forget that a barrel full of water (whether it's 55 gallons or not) is quite heavy, so it's going to stay put until the water has been used. Some people "daisy chain" multiple rain barrels together, putting the overflow pipe from the initial barrel into a second barrel and on down the line.

You can generally count on gravity to help you drain the water out of your rain barrel. However, if you need to move the water to a location that's uphill of the barrel, you'll want to get a small water pump. (These are available through companies that sell rain barrel equipment—see Appendix B.)

Other Options

You don't need a big barrel to harvest rainwater. Even something as small as a 5-gallon bucket can be used to catch rain for watering plants or flushing the toilet. (See the later "Recycling Water" section for more information on forced toilet flushing.) Some home-steaders amass a stockpile of buckets and put one under each downspout when it rains.

Remember that water-filled buckets should be promptly covered to prevent mosquito breeding. Also, while a 5-gallon bucket shouldn't be difficult to handle when full of water, larger buckets can quickly become too heavy to move.

Another (non-roof-dependent) option for harvesting rainwater is to attach plastic sheeting to bushes. Use clothespins to secure the plastic, and fashion the lowest end into a point, under which you can place water jugs to collect the rainwater. You can create a funnel by cutting off the top half of a gallon-size plastic milk jug and placing it upside down in your water receptacle.

SMALL STEPS

Although drinking harvested rainwater isn't recommended, some people do it, especially if they can harvest water using plastic sheeting. If you're considering it, wait until it's rained for 30 minutes before harvesting. That gives the rain time to clear the air.

Tax Credits and Legal Issues

Depending on where you live, tax credits and rebates—or even grants—may be available to help cover the cost of a water harvesting system. Besides the obvious benefit of reducing municipal water usage, uncontrolled storm water runoff can pollute waterways by carrying oil, chemicals, and trash into the local water system. Therefore, many cities and states are encouraging residents to collect rainwater and redirect it for irrigation.

All these benefits are location and situation specific, so you'll need to do some research before moving ahead. Local laws change all the time, so the best way to find out what's currently available is to contact your city's water division.

So that's the good news—rainwater harvesting is encouraged in lots of places. The bad news is that in some states rainwater harvesting is illegal, with some small exceptions. The reasoning behind this is that the water that falls on your property is not really yours to use. It belongs to people down the river who own water rights. In drier parts of the country, there are more people with water rights than actual water, so folks tend to get a little persnickety about it.

However, don't get discouraged if you live in one of these areas. There's a push to reexamine these laws, and in some cases, restrictions are loosening. Plus, homesteaders in those states typically say the powers that be aren't eager to haul them into court over a rain barrel.

You can also reconsider what you call your project. Rainwater "harvesting" or "storing" is not allowed, but often "storm water management" or "redirection" is fine. It seems you're not allowed to hoard water for your own use, including irrigation, but you can slow it down and redirect it to another point on your property. Sound confusing? It is, but it's a reality for some wannabe water harvesters.

The previous paragraph is not intended to be a substitute for legal advice. If you want to know more about your state's water laws, look them up or speak to an attorney.

Recycling Water

It may seem that water is lost to us once it goes down the drain, but there are other options. In the world of water recycling, "white water" is potable (drinkable) water. "Black water" is what comes from toilets and dishwashers and is not considered suitable for reuse. Gray water falls is the middle. It's not potable, but it can be used for some kinds of irrigation. Gray water can be captured from showers, baths, hand sinks, and washing machines, and it constitutes a good portion of a home's wastewater.

ROAD BLOCK

Some gray water advocates consider water that's been used to wash dishes to be usable gray water. However, those with experience in wastewater management state that the food particles in dishwater are attractive to disease-causing organisms. Therefore, dishwater is more biologically similar to black water than gray water.

Gray Water Uses

Although gray water is certainly clean enough to deserve a second life, there are still dangers for bacterial contamination. Because of this, there are limited prescribed uses for gray water.

You can use it for force-flushing your toilet. If you pour a gallon of water into the bowl after using the toilet, it will be forced to flush. This can save a considerable amount of water in the long term, considering toilets comprise about a third of indoor water use. Do not put gray water into the toilet's tank.

Watering nonedibles, like houseplants and landscaping plants, is another good use for gray water. However, gray water should not be used topically on lawns because any bacteria in the water will remain on the surface of the grass. The only way to use gray water for lawns is to bury a drip system under the grass.

Gray water can be used for irrigating some edible plants. It's best to only use gray water on plants where the food is grown well off the ground (like fruit trees) and therefore is not in danger of bacterial contamination.

Don't use gray water for irrigating root vegetables that are not cooked before eating (like carrots and radishes) or any plant in which the edible part is likely to come into direct contact with the water (lettuce, strawberries, etc). It's best to disperse gray water through a drip irrigation system. It should not be used in a sprinkler or sprayed onto foliage.

If you are using recycled water from your shower, hand sink, and washing machine to irrigate landscaping, avoid using any soaps that contain substances harmful to plants. Unfortunately, even some "natural" cleaning supplies (like borax and baking soda) can stress plants. Purchasing soaps labeled "biodegradable" isn't sufficient; you'll need to find products specifically created to be gray water–compatible (see Appendix B for resources).

SMALL STEPS

Gray water is more alkaline than municipal water, and there's a risk of salt buildup on your plants. It's a good idea to occasionally flush your gray water–irrigated plants with some fresh water.

Of course, if you're using gray water to flush your toilet, you don't have to worry about these concerns. However, one rule that applies to all gray water is that it should be used fairly quickly. Don't allow it to sit around for more than a day or so.

Simple Water-Collection Ideas

There are quick and easy ways to capture gray water without making serious alterations to your plumbing system. These options are often the only ones available to apartment dwellers and renters, who may not have the access or permission to monkey around with the plumbing.

You can collect shower water by keeping a 5-gallon bucket next to your shower and using it to catch the water that flows out of the fixture when you first turn it on (and are waiting for the water to warm up). If you have a shower/bathtub combination, plug the drain before you turn on the shower. Then you can catch all your shower water and recycle it just like bathwater.

You can scoop used bathwater out of the tub into buckets, or you can use a little siphon (available at an auto parts store) to draw the water out of the tub.

You can also disconnect the sink. Detach and redirect the drainpipe that connects the bathroom sink to the sewer system into a bucket instead. This trick should not be used on the kitchen sink, where the wastewater contains food particles.

It may seem like a similarly simple thing to disconnect your washing machine from the sewer line. However, we're talking about a lot more wastewater—between 18 and 40 gallons, depending on your washer. That's a lot of water to store and move. The only practical option if you're hoping to disconnect your washing machine is to channel the water directly outside—either toward the ground or to a rain barrel–type container.

Taking It a Step Further

Some people—with more than a healthy dose of plumbing skills—put in extensive gray water systems. This often involves going under the house, cutting or disconnecting drainage pipes, and using new pipes to reroute the water outside. However, because of health and environmental risks, many cities have either outright banned gray water systems or require an extensive permitting process. The time and expense of putting in a permitted system often seem to negate the benefits of recycling the gray water. That being said, some people choose to adapt their plumbing system on their own.

If you're interested in learning how to make significant changes to your plumbing for the purpose of recycling gray water, you can go to oasisdesign.net for more information.

The Least You Need to Know

- Landscaping, primarily in the form of lawns, is responsible for the majority of household water use.
- Purchasing low-flow appliances and fixtures can go a long way toward reducing water usage.
- Rainwater can be stored and then used for irrigation.
- If you observe some precautions, you can capture and recycle some of your home's wastewater.

Turning Waste into Gold

Chapter 222

In This Chapter

- Simple steps for creating compost
- Making your bin
- Tips for composting indoors
- Other ways to reduce your trash

Nothing should be wasted on the self-sufficient holding.

—John Seymour, author of *The Self-Sufficient Life and How to Live It*

If you grow your own food, there's no bigger missed opportunity than a bag full of table scraps and yard waste headed to the landfill. By learning how to compost, you're not only lessening your trash load, you're creating something that will enrich your soil and feed the plants that feed you.

Unfortunately, the stuff that goes into compost makes up only a portion of a home's waste, so it's important to know how to recycle (or re-home) the rest of it. Some items are easy to recycle, but the more toxic stuff can be challenging.

In this chapter, I go over the basics of composting, even if the only space you have available is a balcony. For those who need to compost inside an apartment, a couple of additional options are available to you. I also discuss how to recycle items the recycling center won't take, and what you can do to lessen the number of things you throw away over the course of your day.

Composting in Limited Space

At its heart, *composting* is a simple and natural process. Of course, that doesn't stop folks from creating all kinds of expensive gadgets and complex instructions meant to help you do it "right."

DEFINITION

Composting is the purposeful decomposition of organic matter into natural fertilizer. The process occurs in an aerobic environment (with air) and is accomplished by microorganisms like bacteria.

It's understandable that the process of converting your food scraps into usable fertilizer can be intimidating for those new to it, and it may seem especially daunting to people who live in an urban area. If you were to try composting in a rural setting, you could just pile everything in some far-off corner of the yard; in the city, you need to guard against odor, insects, and critters.

There are some guidelines you can follow to make composting easier and more effective, and you may end up experimenting a little before you find the method that works best for you. However, there's one fundamental principle for anyone who wants to homestead or live a sustainable lifestyle: when it comes to composting, just do it! To compost, even imperfectly, is better than not composting. Don't let any fancy talk about carbon/nitrogen ratios or ideal pile temperatures stand in your way.

What You Can Compost

As simple as composting is, a veteran gardener might be quick to say that composting is not quite the same as "rotting." Rotting would be when you leave a pile of wet grass clippings or food scraps out to do its thing. The pile eventually decomposes, of course, but it's going to get pretty stinky and nasty in the process. What makes composting different is that it relies on a blend of "brown" (or carbon-rich) materials and "green" (or nitrogen-rich) materials. A pile made up of only browns takes an extraordinarily long time to decompose, and a pile of only greens becomes the aforementioned stinky mess. When combined, however, they work together to produce compost.

The stuff that's used for composting is typically called "organic" matter. This is a little misleading, as it might lead people to believe you can only compost materials that are chemical or pesticide free. However, in this case, "organic" refers to things that were once alive or derived in some way from plants or animals. The following table gives you more things appropriate—and not appropriate—for your compost pile.

Potential Composting Materials

Browns	Greens	Do Not Use
Dried leaves	Most table scraps	Meat
Cardboard (shredded)	Grass clippings[4]	Fish
Newspaper[1] (shredded)	Hay	Dairy
Paper[1] (shredded)	Tea bags	Oils
Sawdust[2]	Coffee grounds	Bones
Wood chips or shavings[3]	Weeds[4] (not gone to seed)	Weeds that have gone to seed
Straw	Manure from rabbits, horses, chickens, goats	Diseased plants
Twigs and bark	Fresh leaves	Manure from dogs, cats, pigs, or humans
Dryer lint	Eggshells	
Corn stalks	Not-yet-dead garden plants	
Pine needles[3]		
Dried, brown garden plants		

1 Regular newspaper and white or brown paper (like paper bags) are fine, but avoid using colored paper or glossy, color newspaper inserts.

2 Avoid wood that's been chemically treated.

3 Pine needles can create an acid pH in compost, so use only in moderation.

4 Do not use grass clippings or weeds that have been sprayed with herbicide.

In some gardening books, much ado is made of carbon and nitrogen ratios. The ideal balance is said to be something around 30:1 (carbon:nitrogen). This doesn't mean you have 30 parts browns for 1 part greens; it means you're supposed to know the C:N ratio of each individual ingredient and build your pile accordingly. For example, grass clippings are 19:1, while paper is 170:1. Obviously, building a compost pile in that way would require lots of math and is rather impractical (to put it nicely). A good basic rule is to use around 3 parts browns to 1 part greens.

Ultimately, your compost pile will let you know if the balance is off. If things don't seem to be breaking down, it might be because you have too many browns. If the pile is getting stinky and mucky, you have too many greens.

ROAD BLOCK

There's disagreement within the gardening community as to whether old tomato plants should be composted. Some people believe including them risks transferring tomato diseases to future crops; others say that if you ensure your plants are healthy, it isn't a problem. This may be an area where it's best to err on the side of caution because compost that harbors tomato disease can ruin an entire season's crop.

Building a Compost Bin (or Not)

Ultimately, any time browns and greens are piled together they will begin composting, so you don't need to buy or make a compost bin if you don't want to. Making a mound (preferably on dirt instead of concrete) and covering it with a tarp is a perfectly legitimate way to compost. However, creating a container of some type keeps things a little neater and also allows you to make more compost by giving support to your pile as it grows vertically.

You can purchase any number of commercial composters with a wide array of features, but there's probably no other area of homesteading where it's so simple to create something for yourself. Compost bins don't even have to be "built," per se; they can be fashioned without using any tools.

If you have the space, the ideal size for a compost bin is 3×3×3 feet. This is the size recommended for "hot" composting, which I discuss in the next section. Don't worry if you don't have enough room for a 3-foot bin. Smaller sizes still get the job done.

Here are some ideas for fashioning a compost bin:

Old trash can: You can use either a plastic or metal can with a lid. Just drill a few holes in the bottom (which is easier with plastic cans, of course) to allow the worms to enter and the excess moisture to escape. If you don't have a lid, you can cover the can with a tarp.

Wire: Hardware cloth ($\frac{1}{2}$- or $\frac{1}{4}$-inch) works best for this, but you can also use chicken wire. Just circle the hardware cloth to make a tube, and attach the ends using twisted wire or bungee hooks. You don't want to permanently connect the ends because you'll need to open the circle to easily turn and shovel out the compost. If you want to get a little fancier, you can cut out four squares and tie them together to form a square bin.

Pallets: Wooden pallets are used in all kinds of places for shipping materials. You can find them at grocery or hardware stores and sometimes even next to residential dumpsters. Pallets are rather heavy and can be burdensome to handle, but they make a sturdy compost bin! Just nail four pallets together in a square, or use long zip ties to attach them at the corners.

Composting Basics

You don't have to do the work of composting. Microorganisms like bacteria and fungi do it for you, usually with some assistance from worms. Your job is to provide an environment that makes it easy for them to do their work.

The first thing to know on your way to successful composting is that materials cut into smallish pieces about an inch or so wide compost more effectively. That's why you'll see recommendations that you shred paper and cardboard before adding them to the compost pile. Other long items, like viny garden plants, can be chopped up with garden shears. Some people like to spread their compost materials over the ground and then run them over with a push mower. The truth is, uncut materials will compost eventually, and if you don't feel like fussing around with cutting everything into pieces, that's okay.

Other ways you can help the microorganisms do their thing have to do with temperature, moisture, air, and where the bin (or pile) is located. Ideally, you'd place your compost bin on bare dirt so the worms can wiggle their way into your pile. If you need to put the bin on concrete, elevate it off the ground if you can to help with air circulation. A pallet works great for this, but you may want to put a piece of hardware cloth under the bin so the compost doesn't fall out the bottom. Compost bins can sit on balconies, but you should probably set everything on a tarp so bits of compost don't end up falling onto your downstairs neighbor's head. Regardless of whether your bin is located on dirt or not, try to put it somewhere that's convenient and easy to access.

SMALL STEPS

If you live in an apartment without a balcony, you may be wondering if you can try traditional composting. It's not impossible, but most people don't do it because of a fear of odors. The worm composting I discuss later in this chapter is perfect for apartments, though!

The decomposition that happens in composting is aerobic, meaning it requires air. However, the compost's need for air has to be balanced with its need for moisture. If a pile is allowed to become too dry, it won't break down, and if it's too wet, it turns to sludge. The happy medium you're looking for is a dampness similar to a wet sponge.

If you live in an especially humid climate, you may not need to keep your compost covered. (Although leaving it uncovered might encourage squirrels and other pests to loot.) In most situations, you'll want to cover the compost with a tarp or the trash can lid. You can give air to your compost by "turning" it periodically. In the olden days, this meant getting out the pitchfork and tossing the pile. You may want to just dump out the contents of the can, or shovel out the bin, onto the ground and scoop everything back in. If you don't have room for that, simply stirring things within the bin works.

Hot and Cold Composting

There are two schools of thought regarding composting, and they basically involve how long it takes the materials to break down. The first method is "hot" composting. If you conscientiously follow certain guidelines, you can expect your pile to quite literally heat up—to as much as 160°F! This accelerates the decomposition, and you can go from raw materials to finished compost in as little as a couple of months. This method of composting is likely to result in an end product that's healthier for the garden because the high temperatures can kill any weed seeds or diseased plant parts that might be lurking in the mixture.

Here's how to succeed with hot composting:

- Use a full bin, between 3×3×3 feet and 5×5×5 feet in size.

- Have the correct balance of carbon to nitrogen—around 3 parts browns to 1 part greens.

- Cut the raw material into 1-inch pieces.

- Turn the pile at least once per week.

- Maintain ideal dampness by watering as needed.

URBAN INFO

Sometimes compost piles need a little extra boost of nitrogen to get cooking. Homesteaders (especially if they're male) have been known to find a relatively simple solution to this problem—they pee on the pile. That gives it a good dose of added nitrogen, and it starts heating up!

"Cold" composting is sometimes called "lazy" composting—with a good-natured wink, of course. People who practice this method operate with the idea that the decomposition will happen eventually, and they're not in a particular rush. Anything you can do in the way of moistening, turning, cutting the materials into pieces, etc. helps the process move a little quicker, but as long as your pile isn't smelling bad or harboring pests, you can leave it alone and let it do its thing. If you're taking a more relaxed approach, it may be several months to a year before your materials break down into compost.

Regardless of whether you're aiming to use the hot or cold method, one trick is helpful if you're composting in the city. Many of composting's potential problems—including smells, flies, and pests—can be attributed to exposed green materials. Any time you add greens (like kitchen scraps) to your pile, you need to cover them with a layer of browns. This has the added benefit of keeping your carbon to nitrogen ratio about where you want it. Keep a bag of browns next to your compost pile so you can easily grab a handful whenever you need it.

Starting Your Pile

If you're beginning your compost pile from scratch without a large stock of materials, getting started is pretty simple. You might want to first put down a sprinkling of browns because they're not likely to stick to the bottom of the bin. Add a layer of greens, and cover that with more browns. Remember, you're aiming for a 3 parts browns to 1 part greens ratio. Give everything a light sprinkling of water, and cover. You can continue adding to your pile as more materials become available, watering and turning the mixture as needed.

At some point, you need to decide if you want to stop adding new materials to your pile and just leave it alone to decompose. People who have enough space sometimes create a second compost bin to hold new materials. This gives the first pile time to break down and become finished compost. When it's done, you can empty it out and begin adding your new materials to the first bin and leave the second one alone to decompose.

This is a wonderful scenario if you have the room, but many urban dwellers are lucky if they have enough space for one compost pile. If this is your situation, just know that the finished compost is likely to be near the bottom of the pile. You can go digging periodically and filter out the finished compost using a homemade compost sifter. Just make a frame by nailing four pieces of wood together and stapling or gluing a piece of $\frac{1}{4}$-inch hardware cloth over the frame. Scoop the mixed compost into the sifter, and shake it around until the finished compost falls through into a bucket or tub.

Finding Compost Materials in the City

The primary purpose of composting is to recycle the waste from your household into soil-building fertilizer. So at least hypothetically, you wouldn't need to venture elsewhere to find raw materials. However, it's very likely that your home's waste production won't meet the specific ratios ideal for composting. (Urban homesteaders are often low on browns.) Plus, carting home someone else's waste to convert it to compost is a benevolent form of recycling.

Here are some ideas if you're looking for browns:

- Newspaper recycling centers.

- Community leaf-drop sites.

- Public postings (signs on bulletin boards, notes on Craigslist or Freecycle) stating that you'll pick up bagged leaves.

- Offer to rake the yard of a neighbor—a good deed and compost, all in one!

- Cardboard boxes from grocery stores.

- White paper from office buildings.

- Wood chips from tree care companies.

- Sawdust from construction projects.

- Offer to take away the straw bales used for decoration in garden centers, schools, businesses, etc. after Halloween has passed.

If you're looking more for greens, try the following:

- Used coffee grounds from your local café.

- A restaurant willing to separate and bag its vegetable prep trimmings for you.

- Rabbit or guinea pig manure from a local pet store. (If it's mixed with pine shavings, that's fine.)

- Grass clippings from your neighbors' lawns—only if they don't use herbicides.

Indoor Composting

If you can't or don't want to have a compost bin outside, you still have ways to produce great compost. Even if you're also doing traditional composting, you may want to try an indoor method as a way of supplementing. The more fertilizer, the better!

Let Worms Do the Work

One popular method of composting indoors is called *vermicomposting* and uses worms to break down your scraps. The product of vermicomposting is *worm castings*, lauded by gardeners and farmers as the most coveted of organic fertilizers. Worm castings contain a wealth of water-soluble nutrients, minerals, and beneficial bacteria. Even though they're an incredibly rich fertilizer, they won't "burn" a plant if they're accidentally overapplied. Because of their nutrient density and time-release qualities, worm castings are a favorite additive for seed starting and potting soil mixtures.

DEFINITION

Vermicomposting is the use of worms to transform organic waste into fertilizer. **Worm castings** are a worm's droppings, or manure.

The best worms to use in vermicomposting are known as "red wrigglers" or "redworms." (Their Latin names are *Eisenia foetida* or *Eisenia Andrei*.) You can sometimes find these at bait shops, although you might have to go to a garden center. Redworms are the best for vermicomposting because they breed rapidly and can tolerate fairly large temperature fluctuations. It's definitely not a good idea to set your worm bin in direct sunlight, and you'll also need to bring it inside if the temperature drops below freezing, but other than that your worms should be fine.

Setting Up and Caring for Your Worm Bin

A plastic tub that holds 10 gallons is great for vermicomposting, but you can do it in as small (or as large) of a container as you'd like. It's best if the container isn't clear because worms don't like the light. You'll need to drill some holes along the sides of your bin—near the top is best—to provide airflow.

Many people advise also drilling holes in the bottom of the bin and elevating it on a couple of bricks or wood scraps. This provides an exit for excess moisture, and it also allows you to capture the nutrient-rich liquid (provided you put something there to catch the drips). If you try this, be sure that you put a wire mesh screen in the base of your bin to keep the worms from sneaking out. However, if you're in a situation where a leaking bin is going to be more of a mess than it's worth, you can skip the drainage holes in the bottom. Just take extra care not to overwater your bin because the worms can drown.

SMALL STEPS

Calcium helps worms reproduce, so it's a great thing to add to your worm bin. Just throw in a few crushed eggshells from time to time, and they'll get the calcium they need.

Fill your bin about one third of the way with some bedding made from soft browns. Shredded newspaper, wood chips, or straw are all good choices; bark or twigs are not. Moisten the bedding to the dampness of a wet sponge. Add your worms, and cover them with a bit of moist bedding. You can now begin to feed the worms your scraps—anything from the browns or greens list in the "What You Can Compost" section of this chapter is suitable (except citrus).

You should add anything from the "greens" category in small amounts only. If you give the worms more greens than they can eat, the waste will rot—smelling bad and attracting flies—instead of getting composted. Also, too much nitrogen-rich waste (like chicken manure) at once can actually heat up your bin to a temperature that's dangerous for the worms. If you find you're having a problem with fruit flies, you can try putting a piece of old carpet on top of the compost in the bin.

The most important part of taking care of a worm bin is maintaining the proper moisture. Water-logged bedding can drown the worms, but if things dry out they will also die quickly. It's good to take a moment to check the bedding whenever you feed the worms. Use a stick or spoon to stir things around, testing the dampness of the bedding a few inches down. Eventually, the worms will eat their way through the scraps you feed them as well as all the bedding materials. If you still have space, you can add fresh bedding and continue on with the composting.

Eventually you're going to want to harvest the worm castings—both to create more room in the bin and to use the wonderful fertilizer for your plants.

SMALL STEPS

Rabbit droppings make wonderful fertilizer on their own, but if they're composted by worms, they can be used in seed starting mixture or compost tea. Some homesteaders simplify matters by placing their wire rabbit cage over an open worm bin, providing the worms with a steady supply of food.

Harvesting the Castings

Worm castings look like black, crumbly soil. You can reach into your bin and scoop them out, but then you have to pick out all the worms and return them to the bin—which could take a while. There are a couple of easier options for separating your worms from their castings:

Hardware cloth: You can entice the worms to move where you want them by luring them with one of their favorite foods—banana peels. Cut a piece of ¼-inch hardware cloth so it fits snugly on top of the compost in your bin. Cover the hardware cloth with some moistened bedding, and top it with a few banana peels. After a few days, the worms will find their way to the banana peels. Then you can lift out the hardware cloth and the worms and remove the finished castings. Refill the bin with fresh moistened bedding and dump the worms back in.

Two bins: Drill holes in the bottom of a second bin, and place it on top of the worm compost. Fill the second bin with fresh, moistened bedding and scraps for the worms. Give the worms a month or so to work their way into the second bin. When they've completed their migration, your second bin becomes the main bin, and you can use all the worm castings left behind in the original bin. Repeat as needed whenever the worms fill up their current bin with castings.

A Bit About Bokashi

Bokashi is another option for those interested in indoor composting. *Bokashi* is a Japanese word that basically means "fermented organic matter." In bokashi, waste doesn't really decay—it ferments. The process involves adding a combination of natural bacteria and yeasts called effective microorganisms (EM). The EM is infused onto a carrier (like rice hulls or wheat bran), which is then sprinkled in with the waste scraps. The waste, which can include meat and dairy scraps, is stored in an airtight container.

Although not a completely odor-free way of composting, the fermenting does produce a different—and many people say more pleasant—odor than sometimes happens with traditional composting.

Bokashi is used widely in Japan as well as other island communities as a way of saving landfill space and improving soil. However, the waste scraps do not completely break down during bokashi composting. This means they can be dug into a garden, but they can't be used for seed-starting mixtures or potting soil. Also, the EM must be added to the bokashi daily to ensure consistent fermentation. It's possible to use some of the fermented compost to inoculate the new bin, but many people find it more convenient to use commercially prepared EM.

Making Compost Tea

Compost "tea" may not sound appetizing to you, but your plants will find it delicious. Compost tea is a great way to distribute nutrients, particularly when seedlings are young. The tea is the result of water interacting with compost, and it's available in two ways. The first is by capturing the liquid that naturally seeps out of compost. This can be done with a compost pile, especially if you're using a commercial composter that has a compost tea catcher, but it's especially easy when vermicomposting. Just be sure to drill holes in the bottom of your worm bin, and put a tray under the bin. All the liquid that leaks out is usable compost tea.

The second option is to create your own tea by steeping compost in water. You can use either regular finished compost or worm castings. Just select something to hold the water—it can be as small as a gallon pitcher or as big as a 10-gallon bin. Set aside a proportional amount of compost (between ½ cup and a shovelful). Make a "tea bag" by placing the compost in a piece of cotton fabric or old pantyhose and tying it closed with a string. Steep the compost in the water for around a week, and you'll have concentrated liquid fertilizer.

To use compost tea, dilute the liquid until it's about the color of weak tea and apply it to seedlings and plants.

Recycling Odds and Ends

It's possible to compost an impressive amount of waste, but we know a lot gets left out of the pile. You can find recycling drop-offs in pretty much every city, and many places now offer curbside recycling. Whoever's taking away your recycling can provide you with guides for the various plastics, paper, glass, and aluminum that can be recycled.

The Tricky Items

Some things are notoriously difficult to recycle. This is another area of homesteading where city living has its advantages because there are options nearby for recycling or safely disposing of these items … if you know where to look. You can go to earth911.com to find recycling centers in your area that handle potentially toxic items like batteries, CFLs, and paint. The following table offers some more options.

Recycling Options

Item	Where It Can Be Recycled or Safely Disposed Of
Electronic batteries	Hardware stores, electronics retailers
CFL bulbs	Hardware stores
Car batteries	Auto parts stores
Paint (in buckets)	Your county Household Hazardous Waste center
Old electronics	Used electronics/repair stores
Cell phones	Donate to domestic violence shelter

Item	Where It Can Be Recycled or Safely Disposed Of
Outdated medications	Pharmacy
Motor oil or tires	Auto service providers (usually charge a fee)

There are a few additional options for urban homesteaders looking to limit what ends up in the trash. Both Freecycle (freecycle.org) and Craigslist (craigslist.org) are effective ways to find homes for all kinds of odds and ends. Freecycle is a location-specific mailing list where everything is, of course, free. Craigslist is a message board where things can be either sold or given away. Be aware that Freecycle does not allow "first-come, first-served"–style giveaways; they want you to arrange for pickup of each item with a specific person. So if you're contemplating a big come-and-get-it blowout, Craigslist may be the better choice.

Lastly, centers devoted to hard-to-recycle items are popping up in some communities. They charge a fee for drop-offs, but it can be a way to recycle things like old fabrics or hard plastics. Search online for "hard to recycle (your city)" to see if there's something in your area.

Reducing the Junk

The purpose of homesteading is not to make you (or everyone around you) feel guilty. It's not necessary to live perfectly or to spend your existence in a hut with only handmade tools. However, it is important to recognize that many aspects of our "modern" life are simply unsustainable. We need to be honest with ourselves about our habits and tendencies, one of which is a pretty hefty reliance on disposable items. During a fast-paced day, it can seem easier to walk into the coffee shop empty-handed and throw away the cup when we leave. However, with just a little planning, you can take steps to reduce the amount of waste that passes through your fingers each day.

URBAN INFO

It takes 24 gallons of water to make 1 pound of plastic.

A lot of waste can be eliminated if we simply bring things from home instead of using disposable versions. Here are a few things you can bring of your own:

- Grocery sack to the store
- Coffee cup to the café

- Silverware, plate, and cloth napkin to potlucks

- Food containers (or aluminum foil) to restaurants for taking home leftovers

- Cloth, mesh, or plastic bags to the farmers' market or store for holding greens

- Containers or bags when buying bulk goods (like rice and beans) from the health food store

- Empty water bottle to the airport, so you can fill up after going through security

SMALL STEPS

If you're looking for a silverware option that's lightweight, doesn't take a lot of space, and is easy to clean, try packing a pair of chopsticks!

Reusable water bottles are a great idea in general because it's so energy-inefficient to ship bottles of water across the country. If you like filtered water, purchase a filter that attaches to your kitchen tap—you'll easily save money in the long run. You'll also find that you create less waste—and eat healthier, too—if you pack your own snacks when you go out to movies or a ballgame.

All too often, we take steps to reduce our waste, only to open our mailbox and find a bunch of junk delivered to our door. Ecocycle has created an easy-to-follow list of 10 steps you can take to reduce junk mail at ecocycle.org/junkmail.

Telephone books are another persistent source of unnecessary waste. Call your local phone company and ask that it stop delivering to you. Sometimes phone books are created by independent companies, so you may have to do some sleuthing if you're still receiving books after asking your phone company to stop. If you're not able to stop them, at least the paper can be composted!

The Least You Need to Know

- Composting is not complicated. Really!
- You can create compost in a small yard, on a balcony, or even inside an apartment.
- Composting can be done without a bin, but you can easily make one yourself.
- A lot of the trash we produce every day could be avoided if we brought what we needed from home.

Foraging in the City

In This Chapter

- Picking all the fruit you could want
- Reconsidering weeds
- When trash is actually treasure
- Harvesting fish from city waters

Although we may not always be aware of it, we live in a sea of abundance. Our cities contain a wealth of available food in the trees, dirt, and water. In addition to what nature provides, our fellow citizens regularly discard enough food and goods to support anyone who wishes to take advantage of it (and lots of people do). For some urban homesteaders, the desire to wisely use resources and reduce waste may naturally lead them to forage some of what they eat and use.

In this chapter, I discuss the many possibilities for free food and goods you can find for yourself in a city. Be sure to take note of the legal and safety information included in each section. All the types of foraging I describe here are legal in one form or another, but there are things you'll need to know to participate in a responsible way.

The Bounty of Fruit Trees

From midsummer through fall, there are areas within cities where food is literally hanging from trees, ripe for the picking. Once you begin to notice fruit trees, you'll spot them everywhere, on both private and public land. Most of the time, although not always, the trees on private property weren't planted by the homeowner. They're a relic from an older time—possibly when more people had an interest in harvesting and preserving fruit.

Nowadays much of what grows on urban fruit trees falls to the ground and rots—or gets hijacked by squirrels. However, these underutilized trees present a wealth of possibilities to an industrious urban forager. You may not have the space to plant your own trees or want to wait the many years it takes for a new tree to get established, but you can still harvest a lot of fruit within the city.

Asking Permission

Technically speaking, any fruit trees planted in a right-of-way (the strip of land between the sidewalk and the street) can be legally harvested, and tree branches that hang into alleys or other public walkways are also up for grabs. However, it's always—unequivocally—better to ask.

Before investing too much time, do what you can to examine the tree and be sure the fruit is what you're looking for. Then knock on the front door and talk to the owners, or leave a note if they're not around. Taking a moment to chat also gives you the opportunity to ask whether any chemicals have been sprayed in the vicinity of the tree.

URBAN INFO

An established, city-size apple tree can produce more than 100 pounds of apples per year!

It's likely that when you talk to the homeowners you'll encounter one of two scenarios. Either they've already picked everything they can use, or they don't want any of the fruit at all. In both cases, the homeowners are looking at a future cleanup because of fallen fruit, which means they're probably more than happy to have you take it off their hands. Sometimes the homeowners will even pick some of the fruit themselves and give it to you, just so they know it will be used instead of wasted.

Picking from trees in city-owned parks is a bit of a different story. The rules vary depending on where you live. In some places, groups of people hold publicized coordinated picking sessions. These are presumably not conducted under cover of darkness, so the cities appear to be fine with it. However, some places discourage or outright forbid foraging because of a variety of liability issues.

If you're concerned about a potential foraging bust on your rap sheet, it's best to consult your local ordinances before picking. Sometimes paging through city laws can be a bit too cumbersome to bear. An alternative is to connect with a respected local/sustainable food organization and question some of the veteran members about your city's ordinances.

When You Pick

A great way to grab those fruits just out of reach is to use a fruit picking tool. It's a simple contraption that consists of a long wooden pole with a narrow basket on the end. The wire prongs pull off the fruit, which land in the padded basket. You can purchase fruit pickers online or at garden centers for around $20.

If you're really tempted by fruit that's in the higher branches beyond what a fruit picker can reach, bring a stable stepstool with you. It is not advisable to go climbing into the tree for a couple of reasons. Homeowners and city officials worry about liability if a forager were to fall out of the tree, so it's more respectful to keep your feet on the ground and save them the headache. Also, by climbing, you run the very real risk of breaking tree branches and limbs. That's not good for the long-term health of the tree and will also upset the tree's owner.

As a general rule, on behalf of all the future foragers in your city, use good manners when you're picking. Don't throw the undesirable fruit on the ground, and do tidy up after yourself before you leave.

Tapping into a Network

If you're ready to start foraging but don't know of any trees in your neighborhood, you may be in luck. Some cities have developed fruit tree maps that indicate the location of trees available for picking. Just do an Internet search for "(your city) fruit tree map" and see what you can find. You can also go to neighborhoodfruit.com to see if any trees are mapped in your area. Another good source of information is your city's local food community. The people in those groups will probably know of some good trees.

URBAN INFO

Fruit isn't the only possibility when you go foraging into trees. Some cities also have edible nut trees, such as hazelnuts and pecans.

They say many hands make light work, and when it comes to fruit foraging, there's always enough to go around … and then some. One of the trends in foraging is to form a group and work together to harvest fruit. The fruit is then divided—half goes home with the pickers, and half is donated to a local food pantry. These kinds of community projects are the soul of urban homesteading, and they embody a new way of living and working together in a city. Group work almost inevitably takes on the air of a quilting bee. Everyone talks while they work, and time flies by!

Edible Wild Plants

Edible wild plants sounds pretty exotic, but it's really just a fancy term for "weeds." However, this kind of rebranding is entirely appropriate because the plants' classification as weeds is arbitrary to begin with. As I discussed in Chapter 8, a weed is simply any plant that's growing where you don't want it to. Weeds aren't inherently bad, nor are they automatically inedible. Many of them have been used as food throughout most of human history, and we've only recently started overlooking them.

URBAN INFO

Dandelions, that persistent pest of a weed, were actually cultivated as food by European immigrants, who brought them along when they settled in North America.

Basic Safety and Legal Issues

The first step in foraging wild edibles is ensuring that the plant you're picking is really what you think it is, and I talk more about that in the next section. Beyond that, your primary safety concern is going to be what is on (or in) the plants. Wild edibles that have been drenched with pesticides or chemical fertilizers are not a good deal health-wise, and you probably want to avoid plants that receive regular doses of dog pee. Also, be mindful of the overall land quality where the plants are growing. See "Determining If the Land Is Safe" in Chapter 6 for information about contaminated land and don't forage from areas that are questionable.

As a general rule, avoid foraging any plants growing in or adjacent to well-maintained sod, like a city park or a golf course. You can bet there's an eye-popping amount of chemicals regularly sprayed on that grass to keep it looking nice. Skip plants that grow along well-traveled walkways because anywhere people go, there's likely to be lots of dogs. You should also avoid plants that grow in highway medians or right next to busy roads. They're going to be coated with a lot of nasty exhaust fumes.

The best place to forage wild edibles is in organic gardens—yours and others'. Most community gardens get rather irritated at the idea of nonmembers harvesting, but if you contact the leaders and explain your intention, they might be happy to let you do some weeding. Likewise for any other gardeners in your area or anyone who you feel confident isn't using chemicals in their yard. Lots of public spaces are appropriate for foraging, as long as they're not high-traffic areas.

The main legal issue when foraging wild edibles is to be sure you're not trespassing. Stick to public land unless you have permission from the property owner. Also, try to tread lightly. Don't tromp through a park's flower bed to get to the weeds you've spotted.

A Few Weeds to Get Started

There's no getting around the most important rule of foraging wild edibles—it's best to learn from an actual person, not a book. While most plants are safe and some of the bad ones may only give you a stomachache, there's risk involved when you're learning to forage. There are the occasional wild plants that can be quite toxic. Your city's botanic garden or nature center probably offers wild edibles walks. One advantage of learning from local experts is that they'll be familiar with the types of edible weeds most common in your area.

 ROAD BLOCK

Mushrooms are a popular foraged item. However, mushroom-hunting absolutely *must* be learned directly from a teacher; most homesteading books refuse to even go into it. Look online and join a mushroom foraging group in your area.

In the meantime, here are a few wild edibles you may find growing close to where you live. For a bounty of photos to help you identify these plants, go to google.com/images and do a search. For best results, search by the Latin name rather than the common name—you're more likely to find precisely what you're looking for.

*Dandelion (*Taraxacum officinale*):* The bane of many gardeners' existence is also the queen of the wild edibles. If you want a good chuckle, take a trip to one of the upscale grocery chain stores. It's likely you'll find organic dandelion greens available for sale in bunches—and they're not cheap, either!

Dandelion leaves are best when they're young; try to get them before the plant flowers. The can be served in salads, thrown into smoothies, or cooked like chard. They do have a fairly bitter flavor, but they're quite good for you. Dandelion greens are renowned for cleansing the liver, and they contain lots of vitamins A and C. Dandelion roots are edible, too. They're often dried and used in teas, but you can also prepare them with other root vegetables.

*Purslane (*Portulaca oleracea*):* This is another wild edible that is appearing in gourmet supermarkets and even made the rounds on lists like "The Ten Best Foods You Aren't Eating" in *Men's Health* magazine. Purslane has rounded, succulent leaves and a subtle citrus flavor when eaten raw. It's a great addition to salads or as a green topping for Mexican dishes like tacos and burritos. Purslane can be sautéed or steamed, but it gets slightly slimy (similar to okra). You can use that texture to your advantage by using it as a thickener in soups.

Amaranth/Pigweed (Amaranthus retroflexus): The leaves of this plant can be used just like spinach—eaten raw or lightly steamed. It's a good trick for a gardener to know because spinach can be notoriously finicky and challenging to grow. The plant may be covered with a white powdery substance that has a bitter taste and should be rinsed off before eating. If you'll like, you can also eat the seeds the plant produces after it flowers. However, it may be safer to harvest any amaranth in your garden before it reaches that point because those little seeds spread quickly and produce more amaranth than you know what to do with.

SMALL STEPS

Learning which wild plants can be eaten may cause a shift in the way you approach your garden. You can focus on weeding the inedible plants and harvest the edibles you formerly regarded as troublesome pests. You'll have fewer garden chores and more food!

Chickweed (Stellaria media): This wild plant is high in vitamin C and great in salads. The stems of young plants are edible, too.

Stinging Nettle (Urtica dioica): As the name suggests, this is a plant you should approach with caution. Wear gloves when harvesting, and just clip off the newest growth from the plant. Nettles should not be eaten raw, but once they're steamed or dried, those little stinging hairs are no longer a problem. Cooked nettles can be used like spinach, and dried nettles make a nourishing tea.

Going Diving

The United States is one of the wealthiest nations in the world. Unfortunately, with this great wealth comes great waste. According to a study at the University of Arizona, 40 to 50 percent of the food produced in the United States goes to waste.

The waste is a double-edged sword. Not only does it represent a loss of the natural resources that went into producing the trashed items, but everything that's casually discarded is piling up in landfills.

This glut has given rise to a new group of people, sometimes called *freegans*. Freegans aim to make their way in the world while spending as little money as possible. Instead, they reclaim the food, clothing, electronics, and endless other items that are still usable but thrown away.

DEFINITION

A **freegan** is someone who utilizes discarded goods with the intention of minimizing their consumption of resources.

Freegans embody one possible response to our society's tendencies toward waste, but you can successfully forage from this bounty to whatever extent is comfortable for you. The practice that's somewhat inelegantly known as "dumpster diving" can be a way to acquire free food and goods while simultaneously reducing the burden on our landfills. While dumpster diving may conjure an image of a person launching themselves headfirst into garbage, most trash receptacles can be picked through while standing safely and comfortably on the ground.

Legal Issues

The act of dumpster diving itself is not illegal. In the 1988 case of *California* v. *Greenwood*, the Supreme Court ruled that once an item is discarded, there's no longer a reasonable expectation of privacy or continued ownership. However, there are still a few issues to consider.

Many dumpsters are located on private property, and you can be charged with trespassing for entering without permission. Be mindful of locks and "No Trespassing" signs, even if the dumpster in question is on public property. Businesses often do everything they can to prevent dumpster diving because of a fear of liability if someone gets hurt while in their dumpster.

What You Can Find

If you decide to forage from dumpsters on public land—or even better, if you get permission from the dumpster owner—you can find a lot of good things. Here are a few examples:

Grocery stores: Slightly overripe or bruised produce that is still edible.

SMALL STEPS

If you own backyard livestock, foraged food like produce and bread can be a great supplemental source of feed. Be sure you don't give your animals anything rotten, of course. They still should eat some prepared feed so you can be confident they're getting the right nutrients.

Bakeries: You may find day-old bread, pastries, and other goods still in packaging.

Retail stores: The period of time right after a holiday is a great opportunity to forage all kinds of stuff, from seasonally themed housewares supplies to clothes to candy.

Clothing stores: The change of seasons means a lot of clothes may be thrown away; less-than-perfect returns may also go into the dumpster.

Electronic stores: The fast pace of technological development means there are always newer items to push the old stuff off the shelves. Sometimes it's easier for the stores to dump old inventory rather than sending it somewhere else.

College dormitories: The dumpsters or trash areas around these are freegan goldmines at the end of every school semester.

Rescuing Food Before It Reaches the Dumpster

Restaurants, bakeries, and other establishments work hard to prepare food for their customers, but a lot of it ends up going to waste. Once closing time arrives, much of the day's food heads for the garbage, even if it's still edible. This waste is a commonly accepted practice; it's just factored into the cost of doing business.

SMALL STEPS

Here's another option for feeding your backyard livestock: make arrangements with a local restaurant to periodically pick up its produce trimmings. Use these as supplemental feed for your animals. Be sure to bring the kitchen staff fresh eggs periodically to thank them!

If you're bothered by this waste of food, there's a way to turn the situation into something beneficial. A growing number of people are working with nonprofit agencies to redirect the edible-but-doomed food to needy families. The donating establishments don't need to worry about liability because the Good Samaritan Food Donation Act provides them with protection. You may be able to volunteer with an agency in your area already doing this kind of work. If nothing like this exists yet in your community, you can take steps to pair a restaurant with a food pantry.

Gone (Urban) Fishin'

Fishing may be typically thought of as a pastoral activity, but a surprising amount of it goes on in cities. For example, New York City's East River, lower Hudson River, and Jamaica Bay are home to a wealth of bluefish and striped bass. Many urban lakes and rivers in New York and elsewhere used to be too toxic for fishing. However, over the past few decades, intensive cleanup efforts have resulted in water that can host edible fish. *Forbes* magazine even published a list of the 10 best cities for urban fishing (see Appendix B for the link to the article).

 ROAD BLOCK

Women who are pregnant and children under the age of 15 are especially sensitive to the industrial chemicals that may be present in fish caught in urban waters.

If you want to go fishing in your city, it pays to get to know your state's Division of Wildlife. They can tell you if a lake or river in question is safe for fishing, although many places will have signs posted advising you of their fishing status. You'll also need to purchase a fishing license in order to legally harvest fish from the public waters. An annual license should cost you around $26. The good news is that you're likely to find an old fishing rod on Freecycle or Craigslist. Plus, you can raise your own worms for bait using the vermicomposting techniques discussed in Chapter 22!

The Least You Need to Know

- Most trees, weeds, and dumpsters located on public land can be legally foraged.
- If in doubt, ask permission.
- If you're willing to pick through the trash, you can find a lot that's usable.
- A surprising number of city lakes and rivers contain edible fish.

Glossary

accessory building   A shelter for people, animals, or property that's secondary to the main building on the lot. Greenhouses and chicken coops are examples of accessory buildings.

accessory structure   Something that's separate from, but associated with, the main building on the lot. This includes all accessory buildings but can also refer to structures like a fence.

accessory use   An activity on your property other than what it's zoned to be. For example, if your home is zoned residential, a home-based business or agriculture activities would be an accessory use.

aged manure   Manure that's been out of the animal long enough to be safely applied to plants. Do not use manure from cows or horses that's less than six months old. Chicken manure is best after three months. Goat and rabbit manure do not need to be aged.

annual   A plant that completes its lifecycle within a one-year period.

B.t. (Bacillus thuringiensis)   A bacterium that can be sprinkled on plants for pest management. It kills caterpillars and beetles by exploding their stomach when they eat it. It's safe for humans, but it kills butterfly larvae.

bantam   Small breeds of chicken, weighing between 1 and 2 pounds.

biodiesel   A vegetable oil–based diesel fuel that can be used to power engines. It can also be made with animal fat.

biointensive gardening   A method of gardening wherein the goal is producing as much food as possible from a small space while also improving the soil.

blanch   The process of quickly submerging food in steam or boiling water, followed by rapid cooling in an ice-water bath.

bokashi   A Japanese word meaning "fermented organic matter." This method of indoor composting uses a commercially available starter called effective microorganisms (EM).

bolting The act of a plant sending up a flower stem for the purpose of reproduction. Most vegetables are no longer good for eating after they've bolted.

borax A natural mineral compound used in natural cleaning products and as an emulsifier in lotions.

brine A mixture of salt and water used to preserve food. It may also contain spices, sugar, and/or vinegar.

brooder A heated enclosure used for raising chicks.

brownfield Land that's contaminated with low levels of hazardous waste or pollution.

browns Carbon-rich materials used in composting, like dried leaves and shredded paper.

buck An adult, unneutered male goat or rabbit.

buffer A strip of land that separates two (or more) lots.

builder's sand A coarse sand used as a soil amendment to improve drainage.

cabrito Goat kid meat.

CAE (Caprine Arthritis Encephalitis) Found in goats, this disease is a severe and contagious form of arthritis that results in death.

chevon Goat meat.

coccidiosis A parasitic disease that affects chickens. It can be caused by a buildup of manure and is the most common cause of death in young chicks.

coconut coir Fibers found between inner and outer shells of a coconut. They can be used in place of peat moss in potting soil mixtures.

cold frame A structure placed on top of a garden bed to trap and amplify the sun's heat, which allows food to be grown in the garden bed during cold months. Cold frames typically have a glass panel to conduct the sunshine, but they may also be made with plastic sheeting.

colony collapse disorder (CCD) A phenomenon in which the adult worker bees—or sometimes the entire colony—in a hive abruptly disappear.

colostrum The thick milk a doe produces immediately after giving birth. It's rich in nutrients and antibodies, and without it, a newborn kid has a low chance of survival.

comb The fleshy growth on the top of a chicken's head.

community garden A large piece of land that's divided into small garden plots.

companion planting The practice of strategically positioning plants in combinations that assist with nutrient uptake, pest control, disease management, and pollination.

compost Natural fertilizer that's the result of the decomposition of organic materials derived from plant and animal matter.

composting The purposeful decomposition of organic matter into natural fertilizer. The process occurs in an aerobic environment (with air) and is accomplished by microorganisms like bacteria.

coop A predator-proof enclosure for housing chickens.

cover crop A plant that's grown primarily to add nutrients to the soil. Cover crops are typically tilled or dug into the ground after a specific period of time so they can be incorporated into the soil.

culling The process of removing animals from a group based on specific criteria. Culled animals are typically killed and usually eaten, unless they're removed from the group due to illness.

curds The solids formed during coagulation of milk. Curds contain the milk's fat and most of its protein.

determinate A plant variety that grows to a fixed size and ripens all its fruits at around the same time.

doe An adult female goat or rabbit.

double-dig A method of preparing garden beds for planting. The basic technique is to remove the top layer of soil, loosen the lower layer of soil while digging in soil amendments, and add the top layer of soil back into the bed.

drone bees The male bees in a hive. Their sole responsibility is mating with the queen.

dual-purpose breed A chicken that can be raised either for eggs or meat.

edible landscaping Planting an outdoor space, like a front yard, with vegetables, herbs, and/or fruit-bearing plants like vines, canes, or trees.

emulsifier An ingredient used in lotion to bind together the oil and water elements.

factory farming A term used to describe the practice of raising large quantities of livestock in close confinement; also known as a concentrated animal feed operation (CAFO). Factory farming is the conventional, industrial model for meat, egg, and dairy production.

fingerling A young or small fish used to stock aquaponics tanks.

flagging   When a doe waves her tail rapidly back and forth; it's a sign the doe is in heat.

food-producing animals (FPA)   A term commonly used to describe small animals raised in an urban setting for the primary purpose of food production. Chickens, ducks, dwarf goats, rabbits, fish, and bees are FPAs.

freegan   Someone who utilizes discarded goods with the intention of minimizing their consumption of resources.

fry   Very small baby fish.

germination   The process by which a plant emerges from a seed and begins growth.

grandfathering   Allowing existing structures or practices to continue, even if they're not compliant with updated zoning rules.

gray water   Wastewater generated from bathing, showering, hand washing, and laundry that can be reused for certain types of irrigation. Wastewater from toilets and dishwashers is not gray water.

green roof   A roof covered with vegetation and a thin layer of growing medium, which is typically placed over a waterproof shield. Edible plants aren't usually grown in this manner because the growing medium isn't deep enough.

green space   An area covered with some kind of plant growth; this can include roofs.

greens   Nitrogen-rich materials used for composting, like grass clippings and vegetable table scraps.

hardening off   The process of gradually exposing seedlings to outdoor conditions before planting them in a garden or outdoor container.

hardware cloth   A flexible wire mesh material, consisting of wires woven together in a grid.

hay   Grass or other plants like alfalfa or clover that are dried and used to feed ruminant animals, such as goats.

heat (i.e., in heat)   Ovulating and ready to be bred.

heirloom   Plant varieties that are open-pollinated and maintain their traits from one generation to another.

hutch burn   A condition in rabbits that results in red and irritated skin around the genitals. It is the result of ammonia from soiled bedding.

hybrid   Plant varieties that are the result of a cross between two similar varieties. They do not produce seeds that maintain the characteristics of the parent plant.

impervious surface A surface that does not permit the absorption of fluids, like rooftops or patios.

indeterminate A plant variety that continues growing until it's killed by a stressor like frost or disease. It continues to set and ripen fruit throughout the course of its life.

kid A goat less than 1 year old.

kidding The act of a goat giving birth.

kilowatt-hour A unit of energy equivalent to 1,000 watts of power expended for 1 hour. Kilowatt-hours (kWh) are the way household energy use is measured and billed by the utility company.

kindling The act of a rabbit giving birth.

kit A baby rabbit.

land-share agreement A contract that dictates terms when one person will be growing food on another person's land.

lasagna gardening A technique that involves layering browns and greens and leaving them to compost to create a quantity of new soil suitable for planting. This method allows for gardening without digging into the ground.

layer A chicken bred for its egg-laying qualities and raised for egg production.

leaf mold Compost made from decaying leaves.

litter The group of rabbit babies produced in a single birth.

mixed use A zoning designation that allows for more than one classification of activity. For example, some lots are zoned both residential and commercial.

mulch A covering placed on the surface of soil for the purpose of preventing weeds, conserving moisture, and/or regulating soil temperature.

multigraft fruit tree A tree with three or four compatible varieties grafted (attached) to a single tree so the tree will self-pollinate.

nonconforming use A use of a property that's allowed to continue even after a new zoning ordinance prohibiting it has been established for that area. If someone was allowed to keep chickens on their lot under an older zoning code, and their right to keep chickens was grandfathered in, the chickens would be a nonconforming use.

ordinance A local law or regulation.

ornamental A plant grown for its beauty.

pallet A flat structure that supports goods while they're being lifted by large moving equipment like a forklift.

pasteurization The process of heating milk to slow microbial growth and reduce viable pathogens. Milk must be heated to 145°F for 30 minutes or 163°F for 30 seconds. This also refers to the practice of heating or freezing dried foods to prevent contamination by insects or mold.

peat moss Also known as sphagnum moss, peat moss grows in bogs and is harvested for potting soil mixtures, among other uses. Peat moss is not considered a sustainable resource because its removal threatens the health of bogs.

pelt The skin of an animal with the fur still attached.

perennial A plant that lives for more than two years.

perennial herb An herb plant that lives for more than one growing season. Even if the plant appears to die during the winter, it will typically thrive again in the spring. Popular perennial herbs include thyme, oregano, tarragon, and sage.

perlite A naturally occurring mineral used in soil mixtures to prevent water loss and soil compaction.

permaculture A contraction of the words *permanent* and *agriculture*, this refers to the creation of a system that imitates the relationships found in nature.

pH The measure of acidity or alkalinity in soil or another mixture. For the purposes of canning, the pH determines whether a food has enough acid to be safely canned in a water bath or whether it requires pressure canning.

planting frame A grid created by attaching fencing material or chicken wire over a wood frame. It's used as a guide for plant spacing when creating a garden.

pollination The transferring of pollen from one flower to another so the plant will reproduce.

primary use The activity a lot is zoned for, like residing on a lot zoned residential.

propolis A resinous substance bees collect from trees and plants. It's used as a sealant in the construction of their hives and also has antimicrobial properties.

pullet A hen less than 1 year old. Often used to refer to a young hen who's just started laying, which would be around 5 months old.

queen bee A large female bee responsible for laying all the eggs for the hive and generating a mixture of pheromones, which regulates the behavior of the other bees in the colony.

rabbitry   A place where domesticated rabbits are kept and typically bred.

raised beds   Garden beds built above the ground. The bed frame is typically constructed of wood, but can be made from other materials, and is filled with a growing medium before planting.

raw milk   Milk that has not been pasteurized.

rennet   An enzyme that coagulates milk. It can be obtained from animal or plant sources.

repotting   The process of transferring plants from one container to another.

right-of-way   The strip of land located between the sidewalk and the street, also known as a treelawn, verge, or "hell strip." The right-of-way is typically owned by the city, but the resident of the property associated with it must maintain it.

root vegetables   Plants in which the root is eaten, such as beets, carrots, turnips, and potatoes.

roving   A piece of wool that's been washed, carded (combed), and gently twisted together to prepare it for spinning.

royal jelly   A nutritious substance secreted by worker bees and fed to all larvae in a colony. Feeding extra royal jelly causes a larva to develop into a queen bee.

run   A chicken wire–covered enclosure that allows chickens to spend time outdoors while remaining in a contained space.

secondary use   Incidental uses of a property. If a house and a garden share a lot in a residential district, the garden is considered the secondary use.

setback   The distance of a structure from the property line.

soil amendment   Any material added to the soil to improve its physical properties, resulting in better plant growth and health.

solarization   The process of using the heat of the sun to raise soil temperature to a point that kills weed seeds as well as soil-dwelling pests and diseases. This is accomplished by securing plastic over the surface of the soil for a sustained period of time.

spawn   Started mushroom growth (the mushroom equivalent of a vegetable seedling). It can also mean the production of large quantities of eggs in water—the means by which fish and amphibians reproduce.

sprouting   Soaking and rinsing seeds, beans, grains, or nuts until they germinate.

straw   The dried stalks of cereal plants like oats and wheat after the edible parts have been removed. Straw is used for animal bedding.

strip cup   A cup with a wire mesh filter. The first squirts of milk are directed into the strip cup when milking, and the filter shows any abnormalities that may be present in the milk.

succession planting   The practice of planting a garden bed two or more times during a growing season. Plantings are planned in a way that supports healthy plant growth and does not deplete the soil of nutrients.

supers   Part of a Langstroth beehive, these removable upper compartments are where worker bees store honey.

supersede   The process by which a bee colony replaces an old or inferior queen with a new queen.

thinning   The practice of selectively pulling plants out of a crowded garden bed to give the remaining plants more room to grow. The term also applies to removing some of the fruit from a fruit tree to support the health of the tree and the growth of the remaining fruit.

urban homesteading   A collection of practices, which can be done within a city, with the aim of meeting basic daily needs in a self-sufficient and sustainable way. Urban home-steading practices include growing and preserving food, raising animals for food, creating natural cleaning and body care products, using solar energy, and capturing and recycling water.

use by right   A use permitted within a specified zone or district that does not require special review or approval. For example, you can own a hamster as a "use by right," but you might need a permit before you can own a chicken.

variance   Permission to do something that's not allowed under the zoning code. For example, you'd need to get a zoning variance if you wanted to build a home that's higher than what's allowed in your area.

vent   The opening on a chicken's backside through which they emit eggs and droppings.

vermicomposting   The use of worms to transform organic waste into fertilizer.

vermiculite   A naturally occurring mineral used in soil mixtures to prevent water loss and soil compaction.

washing soda   Also known as sodium carbonate, this is used in natural cleaning prod-ucts as a water softener, as a deodorizer, and for removing grease.

water bath canning   A method of preserving high-acid foods with a pH of 4.6 or lower that involves placing the ingredients in canning jars and submerging in boiling water for a specified period of time.

wattles   The flaps of flesh that dangle under a chicken's chin.

wether   A neutered male goat.

whey   The watery part of milk left behind when the curds are removed. It contains milk's water, sugar, minerals, and some protein.

worker bee   Nonreproductive female bees who do the work of maintaining the hive, including cleaning, tending to the queen, and foraging for nectar and pollen.

worm castings   A worm's droppings, or manure.

xeriscaping   Landscaping in ways that reduce or eliminate the need for supplemental irrigation.

zoning code   The set of zoning regulations for your city.

zoning regulations   Rules created by local government regarding the location, size, structure, and function of a property (or property improvement) in a given area.

Resources

Throughout the book, I've mentioned Appendix B as a source of additional information. Each chapter contains references to specific information—like building a goat milking stand—you can find here. Simply look under the chapter heading to find the resource for the skills and information referred to in that chapter. All of the books are listed under the part headings, since the information in one book could be applied to several different chapters.

To find additional information on urban homesteading and sustainable living, including clickable links to the websites in this appendix, go to eatwhereUlive.com.

General Homesteading and Sustainable Living

Edible Communities: ediblecommunities.com

Food, Inc.: foodincmovie.com

Food Politics: foodpolitics.com

King Corn: kingcorn.net

La Vida Locavore: lavidalocavore.org

Living a Simple Life: livingasimplelife.com

Local Harvest: localharvest.org

Meetup (search for homesteading-type groups in your area): meetup.com

Mother Earth News: motherearthnews.com

Slow Food: slowfoodusa.org

Ten Apple Farm: tenapplefarm.com

Transition U.S.: transitionus.org

Belanger, Jerome D. *The Complete Idiot's Guide to Self-Sufficient Living*. Indianapolis: Alpha Books, 2009.

Berry, Wendell. *Bringing It to the Table: On Farming and Food*. Berkeley: Counterpoint, 2009.

Emery, Carla. *The Encyclopedia of Country Living*. Seattle: Sasquatch Books, 2008.

Gumpert, David. *The Raw Milk Revolution: Behind America's Emerging Battle Over Food Rights*. White River Junction: Chelsea Green Publishing Company, 2009.

Kimbrell, Andrew. *Fatal Harvest: The Tragedy of Industrial Agriculture*. Sausalito: Foundation for Deep Ecology, 2002.

Pollan, Michael. *Food Rules: An Eater's Manual*. New York: Penguin, 2009.

———. *The Omnivore's Dilemma: A Natural History of Four Meals*. New York: Penguin, 2006.

Seymour, John. *The Self-Sufficient Life and How to Live It*. New York: Doris Kindersley, 2009.

Inspiration

Brende, Eric. *Better Off: Flipping the Switch on Technology*. New York: HarperCollins, 2005.

Fine, Doug. *Farewell, My Subaru: An Epic Adventure in Local Living*. New York: Villard Books, 2009.

Gussow, Joan Dye. *This Organic Life: Confessions of a Suburban Homesteader*. White River Junction: Chelsea Green Publishing Company, 2001.

Hathaway, Margaret, and Karl Schatz. *The Year of the Goat: 40,000 Miles and the Quest for the Perfect Cheese*. Guilford: The Lyons Press, 2007.

Kessler, Brad. *Goat Song: A Seasonal Life, a Short History of Herding, and the Art of Making Cheese*. New York: Scribner, 2009.

Kingsolver, Barbara. *Small Wonder: Essays*. New York: HarperCollins, 2002. (particularly the essays "A Fist in the Eye of God" and "Lily's Chickens")

Kingsolver, Barbara, Steven L. Hopp, and Camille Kingsolver. *Animal, Vegetable, Miracle: A Year of Food Life*. New York: HarperCollins 2007.

Part 1: What It Means to Homestead in the City

EatWhereULive: eatwhereUlive.com

Homegrown Evolution: homegrownevolution.com

The Urban Conversion: theurbanconversion.com

Carpenter, Novella. *Farm City: The Education of an Urban Farmer.* New York: Penguin, 2009.

Coyne, Kelly, and Erik Knutzen. *The Urban Homestead.* Port Townsend: Process Media, 2010.

Chapter 2

"Right to Dry" campaign: laundrylist.org/en/right-to-dry

Chapter 3

"So You (Don't Particularly) Want to Be a Farmer" essay: http://bit.ly/hEg6RH

Part 2: City Farming

American Community Gardening Association: communitygarden.org

City Farmer: cityfarmer.info

Garden Guides: gardenguides.com

GardenWeb Forum: forums.gardenweb.com/forums/

Guerrilla Gardening: guerrillagardening.org

International Seed Saving Institute: seedsave.org

National Sustainable Agriculture Information Service: attra.org

Organic Gardening: organicgardening.com

Organic Gardening Forum: forums.organicgardening.com/eve

Vertical Veg: verticalveg.org.uk

Ashworth, Suzanne. *Seed to Seed: Seed Saving and Growing Techniques for Vegetable Gardeners.* Decorah: Seed Savers Exchange, 2002.

Bradley, Fern Marshall, Barbara W. Ellis, and Ellen Phillips, eds. *Rodale's Ultimate Encyclopedia of Organic Gardening.* Rodale Press, Inc., 2009.

Coleman, Eliot. *Four-Season Harvest: Organic Vegetables from Your Home Garden All Year Long.* White River Junction: Chelsea Green Publishing Company, 1999.

Ellis, Barabara W., and Fern Marshall Bradley, eds. *The Organic Gardener's Handbook of Natural Insect and Disease Control.* Emmaus: Rodale Press, Inc., 1996.

Flores, Heather C. *Food Not Lawns: How to Turn Your Yard into a Garden and Your Neighborhood into a Community.* White River Junction: Chelsea Green Publishing Company, 2006.

Hemenway, Toby. *Gaia's Garden: A Guide to Home-Scale Permaculture.* White River Junction: Chelsea Green Publishing Company, 2009.

Jeavons, John. *How to Grow More Vegetables: And Fruits, Nuts, Berries, Grains, and Other Crops Than You Ever Thought Possible on Less Land Than You Can Imagine.* Berkeley: Ten Speed Press, 2006.

Mollison, Bill. *Introduction to Permaculture.* Tasmania: Tagari Publications, 1991.

Riotte, Louise. *Carrots Love Tomatoes: Secrets of Companion Planting for Successful Gardening.* North Adams: Storey Publishing, 1998.

Tracey, David. *Guerrilla Gardening: A Manualfesto.* Gabriola Island, BC: New Society Publishers, 2007.

Chapter 7

Build a PVC cold frame: pvcplans.com/ColdFrame.pdf

Part 3: Raising Animals for Food

Aquaponic Gardening Forum: aquaponicscommunity.com

The Aquaponic Source: theaquaponicsource.com

Backyard Aquaponics: backyardaquaponics.com

Backyard Chickens Forum: backyardchickens.com

BackYard Hive: backyardhive.com

Bee Landing: beelanding.com

Bush Farms (natural beekeeping): bushfarms.com

Fiasco Farm (goat resources): fiascofarm.com

The Goat Spot Forum: thegoatspot.nct

Growing Power: growingpower.org

Murray McMurray Hatchery: mcmurrayhatchery.com

My Pet Chicken: mypetchicken.com

Practical Aquaponics: aquaponics.net.au

Belanger, Jerome D. *The Complete Idiot's Guide to Raising Chickens.* Indianapolis: Alpha Books, 2010.

———. *Storey's Guide to Raising Dairy Goats.* North Adams: Storey Publishing 2001.

Bennett, Bob. *Storey's Guide to Raising Rabbits.* North Adams: Storey Publishing, 2009.

Bernstein, Sylvia. *Aquaponic Gardening: A Step by Step Guide to Growing Vegetables and Fish Together.* Gabriola Island, BC: New Society Publishers, 2011.

Damerow, Gail. *The Chicken Health Handbook.* North Adams: Storey Publishing, 1994.

———. *Storey's Guide to Raising Chickens.* North Adams: Storey Publishing, 1995.

Hathaway, Margaret. *Living with Goats: Everything You Need to Know to Raise Your Own Backyard Herd.* Guilford: The Lyons Press, 2010.

Nelson, Rebecca L. *Aquaponic Food Production.* Montello: Nelson and Pade, Inc., 2008.

Stiglitz, Dean, and Laurie Herboldsheimer. *The Complete Idiot's Guide to Beekeeping.* Indianapolis: Alpha Books, 2010.

Chapter 11

Chicken forums: backyardchickens.com, mypetchicken.com

Coop designs: backyardchickens.com/coopdesigns.html

Coop plans: back-yard-chicken-coop.com

Chapter 12

Herbal wormer and hoof trimmer: hoeggergoatsupply.com

The Goat Spot message board: thegoatspot.net

Milk stand plans: fiascofarm.com/goats/milkstand.html; *Storey's Guide to Raising Dairy Goats*

Photo slideshows of goat births: http://bit.ly/gC7YP8

Chapter 13

Make your own cage/hutch: *Storey's Guide to Raising Rabbits*

Processing a rabbit: *Storey's Guide to Raising Rabbits*; http://bit.ly/h0jnPt

Chapter 14

Build a Langstroth hive: bees-on-the-net.com/langstroth-hive.html

Build a top-bar hive: biobees.com/images/build_top_bar_hive/

Chapter 15

Small tilapia orders: rdaquafarms.com

Part 4: A Homemade Life in the City

Cheeseslave: cheeseslave.com

National Center for Home Food Preservation: uga.edu/nchfp

USDA Complete Guide to Home Canning: foodsaving.com/canning_guide

The Weston A. Price Foundation: westonaprice.org

Bone, Eugenia. *Well-Preserved: Recipes and Techniques for Putting Up Small Batches of Seasonal Food.* New York: Random House 2009.

Briggs, Raleigh. *Make Your Place: Affordable, Sustainable Nesting Skills.* Bloomington: Microcosm Publishing 2009.

Bubel, Mike, and Nancy Bubel. *Root Cellaring: The Simple No-Processing Way to Store Fruits and Vegetables.* North Adams: Storey Publishing, 1991.

Carroll, Ricki. *Home Cheese Making: Recipes for 75 Delicious Cheeses.* North Adams: Storey Publishing, 2002.

Chadwick, Janet. *The Busy Person's Guide to Preserving Food*. North Adams: Storey Publishing, 1995.

Fallon, Sally. *Nourishing Traditions: The Cookbook That Challenges Politically Correct Nutrition and the Diet Dictocrats*. Washington, D.C.: NewTrends Publishing, 2001.

Kalinchuk, Amy. *Making Soap in Your Own Kitchen*. e-book: crafte-revolution.com.

Kingry, Judi, and Lauren Devine. *Ball Complete Book of Home Preserving*. Toronto: Robert Rose Inc., 2006.

Planck, Nina. *Real Food: What to Eat and Why*. New York: Bloomsbury USA, 2006.

Ziedrich, Linda. *The Joy of Pickling*. Boston: The Harvard Common Press 2009.

Chapter 16

Build a solar dryer: builditsolar.com/Projects/Cooking/cooking.htm (click on "food drying")

Build a root cellar: motherearthnews.com (search "basement root cellar")

Chapter 17

Cheese-making supplies: cheesemaking.com

Build a cheese press: fiascofarm.com/dairy/cheesepress.html

Chapter 18

Instructional video for spinning on a drop spindle: http://bit.ly/i29QX9

Chapter 19

Make castille soap: crafte-revolution.com (Making Liquid Soap in Your Own Kitchen)

Part 5: Making the Most of What You Have

Build It Solar: builditsolar.com

Fallen Fruit: fallenfruit.org

Find Solar: findsolar.com

Neighborhood Fruit: neighborhoodfruit.com

Baker, Jerry. *Supermarket Super Gardens.* Wixom: American Master Products, Inc., 2008.

Gibbons, Euell. *Stalking the Wild Asparagus.* Chambersburg: Alan C. Hood, 1962.

Hoffman, John. *The Art and Science of Dumpster Diving.* Port Townsend: Loompanics Unlimited, 1992.

Jenkins, Joseph. *The Humanure Handbook.* Grove City: Jenkins Publishing, 1999,

Kellogg, Scott, and Stacy Pettigrew. *Toolbox for Sustainable City Living.* Brooklyn: South End Press, 2008.

Nyerges, Christopher. *The Self-Sufficient Home: Going Green and Saving Money.* Mechanicsburg: Stackpole Books, 2009.

Schaeffer, John. *Real Goods Solar Living Source Book.* Gaiam Real Goods, 2007.

Thayer, Sam. *The Forager's Harvest: A Guide to Identifying, Harvesting, and Preparing Edible Wild Plants.* Cleveland, NY: The Forager Press, 2006.

Chapter 20

Tiny house video tour: http://bit.ly/hF9SQ6

Build a solar space heater: motherearthnews.com/Do-It-Yourself/1977-09-01/Mothers-Heat-Grabber.aspx

Solar space and water heaters: builditsolar.com

Build a solar cooker: solarcooking.org/plans

Bikes with extra storage: workcycles.com

Chapter 21

Make a rain barrel: epa.gov/region3/p2/make-rainbarrel.pdf

Commercial rain barrel equipment: aquabarrel.com

Gray water–compatible cleaning products: bio-pac.com

Chapter 23

Article on urban fishing spots: http://bit.ly/gGUmQv

Garden Planning Guides

Even the most enthusiastic gardener can feel a little hesitation when it comes time to plan for the spring. You've decided where your garden will be, but there are so many different ways to organize that patch of dirt. To help you with this task, I've provided you with a handy chart of plant information, plus a sample garden plan.

Plant Information Chart

As fascinating as it can be to learn about growing food, it's an awful lot of information to try to hold in your head at once. Therefore, I've created this chart for you to use as a reference. Although the plant spacings may seem radical at first glance, they're based on the practices of a couple of different small space gardening techniques—and have been borne out through my own experience growing food in the city. See the notes following the chart for additional information.

	Direct/ Transplant	Start Seeds Inside	Germination Time	Seedling Growth	Earliest Outside
Asparagus	Transplant crowns	—	—	—	6 weeks before
Arugula	Direct*	—	<7 days	—	Soil thaw
Basil	Either	3 weeks before	7 to 14 days	4 weeks	1 week after
Beans	Direct	—	7 to 14 days	—	1 week after
Beets	Direct	—	10 to 14 days	—	Soil thaw
Broccoli	Transplant	7 weeks before	7 to 10 days	5 weeks	2 weeks before
Cabbage	Transplant	9 weeks before	10 to 14 days	5 weeks	4 weeks before
Carrots	Direct	—	10 to 14 days	—	Soil thaw
Cauliflower	Transplant	7 weeks before	7 to 10 days	5 weeks	2 weeks before
Chard	Direct*	—	10 to 14 days	—	Soil thaw
Collards	Transplant	8 weeks before	7 to 10 days	4 weeks	4 weeks before
Corn	Direct*	—	7 to 10 days	—	1 week after
Cucumbers	Transplant	3 weeks before	<7 days	4 weeks	1 week after
Eggplant	Transplant	9 weeks before	7 to 10 days	10 weeks	1 week after
Garlic	Direct	—	—	—	Fall, before the frost
Herbs (perennial)	Transplant	8 weeks before	14+ days	4 weeks	4 weeks before
Kale	Direct*	—	7 to 10 days	—	Soil thaw
Lettuce	Direct*	—	<7 days	—	Soil thaw
Melons	Transplant	2 weeks before	<7 days	4 weeks	2 weeks after

Time to Maturation	Spacing (Inches)	Square Feet/ Plant	Plant Family	Loves	Hates
Perennial	12	1	Lily	Tomatoes, parsley	None listed
Short	4	0.111	Cole	Carrots, tomatoes	Kale
Long	6	0.25	Mint	Tomatoes, eggplant	Rue
Long	6	0.25	Pea	Cucumbers, corn, carrots	Onions, fennel
Medium	4	0.111	Beet	Bush beans, onions	Pole beans
Medium	15	1.6	Cole	Dill, sage, mint, rosemary	Tomatoes, pole beans
Medium	15	1.6	Cole	Dill, sage, mint, onions	Tomatoes, pole beans
Medium	3	0.0625	Parsley	Tomatoes, onions	Apples
Medium	15	1.6	Cole	Dill, sage, mint, rosemary	Tomatoes, pole beans
Short	8	0.44	Beet	Bush beans, onions, cole	Pole beans
Medium	12	1	Cole	Tomatoes, basil	Grapes
Long	15	1.6	Grass	Potatoes, beans, squash	Tomatoes
Long	12	1	Squash	Corn, beans, peas	Potatoes, herbs
Long	18	2	Tobacco	Beans, peppers	Apricots
Fall planting	6	0.25	Onion	Collards	Beans
All season	6	0.25	Parsley or Mint	Tomatoes, carrots, beans	Cucumbers, melons
Short	8	0.44	Cole	Basil, tomatoes, beans	Grapes
Short	6	0.25	Sunflower	Onions, carrots, cucumbers	Broccoli
Long	18	2	Squash	Corn, radishes	Potatoes, herbs

continues

continued

	Direct/ Transplant	Start Seeds Inside	Germination Time	Seedling Growth	Earliest Outside
Onions, Bunching	Transplant	12 weeks before	<7 days	8 weeks	4 weeks before
Onions, Scallions	Direct	—	<7 days	—	Soil thaw
Peas	Direct	—	7 to 10 days		Soil thaw
Peppers	Transplant	7 weeks before	7 to 10 days	8 weeks	1 week after
Potatoes	Direct	—	10 to 14 days	—	4 weeks before
Radishes	Direct	—	<7 days	—	Soil thaw
Spinach	Direct*	—	10 to 14 days	—	Soil thaw
Squash, Summer	Transplant	3 weeks before	>7 days	4 weeks	1 week after
Squash, Winter	Transplant	3 weeks before	>7 days	4 weeks	1 week after
Tomato	Transplant	7 weeks before	7 to 10 days	8 weeks	1 week after
Turnips	Direct	—	>7 days	—	Soil thaw

Time to Maturation	Spacing (Inches)	Square Feet/ Plant	Plant Family	Loves	Hates
Long	4	0.11	Onion	Beets, lettuce	Peas, beans
Short	3	0.06	Onion	Beets, lettuce	Peas, beans
Medium	4	0.11	Pea	Carrots, cucumbers, radishes, corn	Onions, garlic
Long	12	1	Tobacco	Beans, carrots, onions	Black walnuts
Long	12	1	Tobacco	Bush beans, cole, corn	Cucumbers, melons
Short	3	0.06	Cole	Beets, carrots, lettuce	Hyssop
Short	4	0.11	Beet	Eggplant, peas, strawberries	None listed
Long	21	3	Squash	Beans, corn, radishes	Potatoes
Long	21	3	Squash	Beans, corn, radishes	Potatoes
Long	24	4	Tobacco	Onions, basil, carrots, lettuce	Cole, potatoes
Medium	4	0.111	Cole	Peas, other cole crops	None listed

Some notes on the chart:

Direct/Transplant: Direct means the seeds are sowed directly into the ground. Transplant means the plants should be started indoors as seedlings and then transplanted into the garden.

Start Seeds Inside: When to begin the process of growing indoor seedlings, relative to the projected last frost date. You can learn the projected last frost date for your area by searching online or contacting your agriculture extension office.

Germination Time: Time from the planting of seed to the appearance of growth.

Seedling Growth: Amount of time seedlings should be grown inside before transplanting into the garden.

Earliest Outside: The earliest seeds (for direct-seeded plants) or seedlings (for transplants) can be put into the garden, relative to the projected last frost date.

Time to Maturation: The general amount of time it takes the plant to reach maturity. Specific number of days to maturity differs based on climate and time of year, but plants can be classified as taking a "short," "medium," or "long" amount of time to reach maturity. See "Succession Planting" in Chapter 5 for more information.

Spacing (Inches): The amount of space needed between seeds or seedlings.

Square Feet/Plant: The square footage required for each individual plant; useful when determining how many plants can fit into a given area. For example, each basil plant needs 0.25 square feet. If a gardener has 4 square feet to devote to basil, he or she can have 16 plants (4 ÷ 0.25 = 16).

Plant Family: Helpful in companion planting; plants from the same family share similar characteristics.

Loves/Hates: Examples of pairings that work well and those that should be avoided when doing companion planting. For more information about companion plant relationships, read *Carrots Love Tomatoes* by Louise Riotte (Storey Publishing, 1998).

Direct*: These plants typically are sown directly into the soil as seeds but can be started indoors as seedlings and then transplanted. Generally speaking, seedlings should be started a few weeks before the indicated "Earliest Outside" date.

Soil thaw: The earliest the ground can be worked in the spring; whenever the soil thaws enough to dig with a shovel.

All season: Perennial herbs mature quickly but can continue to be harvested throughout the growing season.

Sample Garden Map

There are as many ways to plan a garden as there are gardeners. I've provided one example to demonstrate an efficient way to utilize a 256-square-foot planting space. The garden contains three beds that are 4 feet wide, separated by paths that are 2 feet wide. This configuration yields 192 square feet of plantable space. As you can see, the garden has been divided into chunks, which correlate to the bed plans in the following chart.

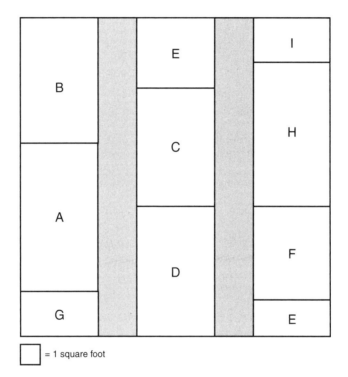

= 1 square foot

Sample Garden Planning Chart

The following chart lays out the planting (and replanting) of each garden bed over the course of the growing season. The dates of the last and first frost will depend on where you live. Contact your local agriculture extension office to get the frost dates for your area.

The last and first frost dates provide a structure for how the garden should be planned. As I indicated in the plant information chart, the frost dates determine when most plants can be put in the garden, and therefore when they should be started from seed.

You can build a chart like this as you plan your garden. Begin by inserting the last and first frost dates. Then, decide which crop you would like to "anchor" each bed during the summer—usually it's tomatoes, peppers, squash, or something like that. Use the "earliest outside" value in the plant information chart to tell when the vegetable should be planted in the garden. If you plan on growing your own seedlings, count back the designated number of weeks (the amount of time required for seedling growth) to tell you when you should start the seedlings inside.

Your ability to successively plant different crops in one bed will depend somewhat on the length of your growing season. However, pretty much everyone should be able to at least sneak in a quick crop (from the "short" time to maturation in the plant information chart) before they put their main summer plants in the ground.

	11	10	9	8	7	6	5	4	3	2	1	Last Frost	1	2	3	4	5	6	7	8	9	10	11	12	13	14	15	16	17	18	19	20	21	Last Frost	1	2	3
A																																					
B																																					
C																																					
D																																					
E																																					
F																																					
G																																					
H																																					
I																																					

Key:

PI	plant seed inside (start as a seedling)
PO	plant seed outside (sowed directly into the ground)
T	transplant seedling into the garden
H	begin harvesting
---	plant is growing indoors as a seedling
———	plant is growing in the garden but is not ready for harvest
===	plant can be harvested

Here's what can be grown in the garden depicted in the Sample Garden Map using these plans:

6 winter squash	172 beets
7 tomatoes	172 peas
10 summer squash	208 lettuce
12 eggplants	263 arugulas
25 peppers	383 radishes
36 perennial herbs	416 scallions
86 turnips	483 radish sprouts
108 kale	490 turnip sprouts/greens
150 carrots	526 spinach

Index

H